Complete Science Work-Book 2

Hobson
Murphy
Wheeler

Scholarstown Educational Publishers

© Scholarstown Educational Publishers, 1987.
Artwork by Michael Phillips
ISBN 1 85276 003 6

All rights reserved. No part of this publication may be reproduced or transmitted in any form or by any means (stencilling, photocopying, etc) for whatever purpose, even purely educational, without written permission from the publishers.

First Published 1987
by Scholarstown Educational Publishers Ltd.
London House, 26-40 Kensington High Street,
London W8 4PF.

Contents

Chemistry

Chapter 1. Atomic Structure 5
Chapter 2. Chemical Bonding 9
Chapter 3. Chemical Formulae 13
Chapter 4. Oxygen 16
Chapter 5. Acids and Bases 21
Chapter 6. Preparation of Salts 24
Chapter 7. Carbon and Carbon Dioxide 27
Chapter 8. Hardness of Water 31
Chapter 9. Oxidation and Reduction 35
Chapter 10. Activity Series 38
Chapter 11. Sulphur and SO_2 42
Chapter 12. Ammonia 45
Chapter 13. Chlorine 49
Chapter 14. Thermochemistry 52
Chapter 15. Calculations in Chemistry 56

Biology

Chapter 16. Food Types 58
Chapter 17. Animal Nutrition 62
Chapter 18. Teeth 65
Chapter 19. Plant Nutrition 68
Chapter 20. Respiration 73
Chapter 21. Water and Transport in Plants 79
Chapter 22. Blood 84
Chapter 23. Excretion in Man 88
Chapter 24. Movement 91
Chapter 25. Sensitivity 94
Chapter 26. Sense Organs 99
Chapter 27. Cycles in Nature 103
Chapter 28. Ecology 106
Chapter 29. Genetics 110

Physics

Chapter 30. Force 113
Chapter 31. Turning Forces 118
Chapter 32. Motion 123
Chapter 33. Pressure 126
Chapter 34. Gas Laws 130
Chapter 35. Heat Transfer 135
Chapter 36. Reflection 139
Chapter 37. Refraction 144
Chapter 38. Optical Instruments 150
Chapter 39. Waves and Sound 155
Chapter 40. Static Electricity 160
Chapter 41. Current Electricity 164
Chapter 42. Effects of Electricity Current 168
Chapter 43. Energy and Heating 174

1 — Atomic Structure and the Periodic Table

Multiple Choice Section

Questions 1-5 concern the following

    A — Proton
    B — Neutron
    C — Electron
    D — Nucleus
    E — Orbits

Choose the BEST answer from A to E

1. The heaviest part of an atom.

2. The positive particle in an atom.

3. The lightest particle inside an atom.

4. The electrons move in.

5. The particle without a charge.

Questions 6-10

For each of these questions ONE or MORE of the answers are correct. Decide which of the responses is (are) correct, then choose.

A	B	C	D	E
1,2,3 correct	1,2 only	2,3 only	1 only	3 only

6. Isotopes have the same number of
   1. neutrons
   2. protons
   3. electrons

7. The family of unreactive elements is known as
   1. alkali metals
   2. halogens
   3. noble gases

8. Which of the following sets of elements are in the same group of the Periodic table?
   1. fluorine, chlorine, bromine.
   2. lithium, sodium, potassium.
   3. oxygen, nitrogen, sulphur.

9. The atomic number of an element is equal to
   1. the number of protons
   2. the number of neutrons
   3. the relative atomic mass of the element

10. A metallic element would
    1. conduct electricity
    2. be shiny
    3. be ductile

Questions 11-15

In the following questions there are five possible answers A, B, C, D and E, choose the BEST answer.

11. The elements in the Periodic table are arranged in order of increasing
    A — relative atomic mass
    B — atomic number
    C — mass number
    D — number of neutrons
    E — number of isotopes

12. Hydrogen (atomic number 1, mass number 1) contains
    A — 1 proton, 1 electron, 1 neutron
    B — 1 proton, 1 electron
    C — 1 electron, 1 neutron.
    D — 1 proton, 1 neutron
    E — 1 proton, 2 neutrons

13. Copper II sulphate is blue. Copper is
    A — an alkali metal
    B — an alkaline earth metal
    C — a halogen
    D — a transition element
    E — a noble gas

14. Which of the following will NOT conduct electricity
    A — graphite
    B — copper
    C — sulphur
    D — iron
    E — magnesium

15. Sodium has the atomic number 11. The electron configuration is
    A — 1.8.2.
    B — 8.2.1
    C — 2.2.6.1
    D — 2.1.6.2
    E — 2.8.1

Structured Section

1.
Bromine	Calcium	Carbon	Chlorine
Copper	Iron	Magnesium	Nitrogen
Sodium	Oxygen	Potassium	Sulphur

(a) Underline all the non-metals in the list

(b) Give the names of all the transition metals in the list. ....
....

(c) Give the names of the halogens in the list. ....
....

(d) Give the names of the alkali metals in the list. ....
....

(e) Give the names of the alkaline earth metals in the list. ....
....

2. In this question you must use the Periodic table shown below. The same letter may be used once, more than once or not at all.

(a) Give the letters of any TWO elements in the same group. ....
....

(b) What is the name of element A? ....
....

(c) Which letter represents a transition metal? ....

(d) Give the letter of an element which is a gas with single atoms in its molecules ...
....

(e) What is the name used for the horizontal row F to T? ....

Essay Section

1. Name the three main particles in an atom. Draw a table to show the charge and mass of each particle.
   The element boron exists in two form $^{10}_{5}B$ and $^{11}_{5}B$ what are these two forms called? What is the mass number of $^{11}_{5}B$ and what is the atomic number of $^{10}_{5}B$? How many of each of the three main particles are in each form of boron? You must show how you reached your answer.

2. In the Periodic table the elements are arranged in vertical and horizontal divisions. What are the general names for these divisions. In the vertical divisions the elements react in a similar way, what is the reason for this? Describe an experiment which you have seen which shows this similarity.

Practical Assessment

Aim: To investigate the properties of the oxides of the third period elements

Apparatus: Test-tubes, test-tube holder, safety mat, safety glasses, distilled water, Bunsen burner, universal indicator, oxides of magnesium, aluminium, silicon, phosphorus and sulphur.

Method

(The properties of phosphorus and sulphur will be demonstrated by your teacher. Sodium oxide and dichlorine oxide are not usually available in school laboratories but you will be given the information for these compounds).

1. Record the appearance of each compound in a table as shown at the end.

2. Place a small amount of one of the oxides in a test-tube and heat. Notice if it melts. Repeat for each oxide.

3. Place a small amount of oxide in a test-tube and add about 5 cm^3 of distilled water. Notice if it dissolves or if it fumes.

4. Test the solution using universal indicator paper or litmus. Record your results.

5. Repeat 3 and 4 for each substance.

Results

Oxides of	Na	Mg	Al	Si	PS	Cl
Appearance Melting point Effect of water Effect of indicator						

(For melting point put either HIGH or LOW)

2 — Chemical bonding and shapes of molecules

Multiple Choice Section

Questions 1-5

Five possible answers A, B, C, D and E are given for each question, choose the one which you think is correct.

1. A charged atom or group of atoms is known as
   A — an electron
   B — an ion
   C — a molecule
   D — a proton
   E — a neutron ☐

2. An electric current is carried through a solution of a salt in water by
   A — protons
   B — electrons
   C — neutrons
   D — ions
   E — molecules ☐

3. The atomic number of calcium is 20. When calcium forms an ion the electron configuration of the calcium ion is
   A — 2.8.8
   B — 2.2.8.8.
   C — 8.2.2.8
   D — 2.8.8.2
   E — 2.8.2.8. ☐

4. The shape of the molecule of methane CH_4 is
   A — linear
   B — V-shaped
   C — tetrahedral
   D — pyramidal
   E — round ☐

5. A carbon dioxide molecule would
   A — have a high melting point
   B — conduct electricity
   C — be a gas
   D — be a solid
   E — have a high boiling point ☐

Questions 6-10

For each of the questions ONE or MORE of the answers are correct. Decide which of the responses is (are) correct. Then choose

A	B	C	D	E
1,2,3 correct	1,2 only	2,3 only	1 only	3 only

6. When an ionic bond is formed
   1. electrons are shared
   2. an atom loses electrons
   3. an atom gains electrons

7. The atomic numbers of F, Ne, and Na are 9, 10 and 11. Which has the electron configuration 2.8?
   1. F
   2. Ne
   3. Na

8. A double bond is formed when
   1. one pair of electrons is shared
   2. two electrons are lost
   3. two pairs of electrons are shared

9. Sodium chloride is ionic. It is likely to
   1. be crystalline
   2. be soluble in water
   3. melt at a low temperature

10. The water molecule is
    1. tetrahedral
    2. covalent
    3. a liquid

Questions 11-15

In the following questions you must decide if they are TRUE or FALSE. If true put T, if false put F.

11. The chlorine molecule is linear.

12. Ionic compounds when molten are good conductors of electricity.

13. The oxide ion has a 2^+ charge.

14. Covalent compounds have high melting points.

15. The ionic bond is strong

Structured Section

1. N, F, Mg, and Al are the symbols of four elements having atomic numbers 7, 9, 12 and 13 respectively.
   (a) Give the electron configurations of the elements

   N.....................   F................

   Mg...................   Al................

(b) (i) When F forms an ion how many electrons will it lose? ....
   (ii) Write down the formula of the ion formed by F ....
(c) (i) When Mg forms an ion how many electrons will it lose? ....
   (ii) Write down the formula of the ion formed by Mg ....
(d) F also forms a molecule containing a covalent bond. Using only the outer electrons show how the covalent bond is formed.

2. Sodium atomic number 11 and chlorine atomic number 17, combine to form an ionic compound, sodium chloride.
   (a) Draw the electron configurations of sodium and chlorine using rings for energy levels.

   Sodium Na   Chlorine Cl

   (b) Using your diagrams in (a) show how the electrons are transferred when sodium chloride is formed.
   (c) Write the formula for the :
      (i) Sodium ion ....
      (ii) Chloride ion ....

Essay Section

1. (a) Name the different types of chemical bonds and explain the difference between them.
   (b) Give an example for each type of bond.
   (c) Give TWO properties for each type of bond.

2. Covalent molecules have definite shapes. Explain why they have definite shapes. Draw the shapes of the following:-
   (a) Methane CH_4
   (b) Ammonia NH_3
   (c) Water H_2O

Practical Assessment

Aim: To find which solutions conduct electricity

Apparatus: Source of direct current, two graphite electrodes, bulb or ammeter, wires, beakers, distilled water, ethanol, dilute ethanoic acid, tetrachloromethane, and aqueous solutions of sodium chloride, copper II sulphate, ammonium hydroxide, dilute hydrochloric acid.

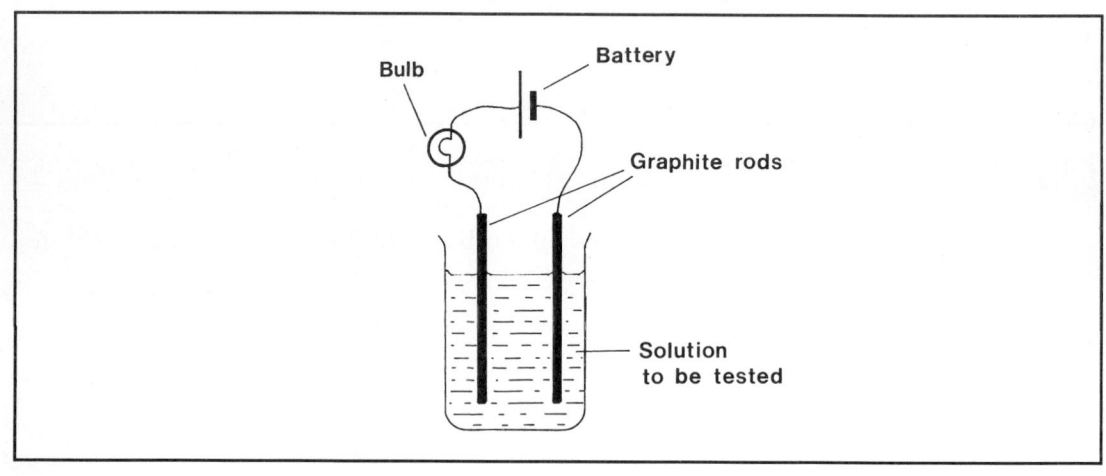

Method

1. Set up the apparatus as shown in the diagram.
2. Half fill a beaker with one of the solutions.
3. Place the graphite electrodes in the beaker.
4. Notice if the bulb lights. If the bulb lights record if it is bright or dim.
5. Record your results in a table.
6. Repeat 2 to 5 for the solutions given, washing out the beaker after each experiment.
7. Note: if using an ammeter record results in Amps.

Complete Science Work-Book 2

3 — Chemical Formulae

Multiple Choice Section

Questions 1-5 concern the following

    A — Valency
    B — Monovalent
    C — Trivalent
    D — Divalent
    E — Radical

Choose the correct answer from A to E.

1. The combining power of an element ☐

2. Oxygen combines with two hydrogen atoms in water because it is ☐

3. The carbonate group is known as a ☐

4. Hydrogen is always ☐

5. Nitrogen combines with three hydrogens in ammonia because it is ☐

Questions 6-10

For each of the questions ONE or MORE of the answers are correct. Decide which of the responses is (are) correct. Then choose:

A	B	C	D	E
1,2,3 correct	1,2 only	2,3 only	1 only	3 only

6. NO^{3-} is known as
   1. a group
   2. a radical
   3. a molecule ☐

7. Copper can have more than one valency, these are
   1. 3
   2. 1
   3. 2 ☐

8. Which of the following equations are balanced?
   1. $C + O_2 \rightarrow CO_2$
   2. $Zn + O_2 \rightarrow ZnO$
   3. $Mg + Cl_2 \rightarrow 2MgCl_2$ ☐

The carbonate group is
1. SO_4^{2-}
2. NO_3^-
3. CO_3^{2-} ☐

10. Balance one or more of the following equations. TWO must be used
    1. $Mg + O_2 \rightarrow MgO$
    2. $Cu + O_2 \rightarrow CuO$
    3. $H_2 + O_2 \rightarrow H_2O$ ☐

Questions 11-15

In the following questions you must decide if they are TRUE or FALSE. If true put T if false F.

11. A chemical equation must always be balanced. ☐

12. When balancing an equation the formulae may be altered. ☐

13. In an ionic compound the charges must be equal. ☐

14. In a covalent compound the valencies of each element must be satisfied. ☐

15. The valency of elements in Group I of the Periodic table is one. ☐

Structured Section

1. There are three oxides of nitrogen having the formulae:-

       NO N_2O NO_2
       (A)   (B)   (C)

   (a) How many nitrogen atoms are in one molecule of NO_2?......................

   (b) Which oxide of nitrogen contains the highest proportion of nirogen?.........

   (c) When oxide (A) is passed over heated copper, copper II oxide and nitrogen are formed. Write the equation for this reaction........................
   ....

   (d) Oxide A combines with oxygen to form oxide C. Write the equation for this reaction ....

2. The chemical equation for the reaction of concentrated hydrochloric acid with manganese IV oxide is:

       $MnO_{2(s)} + 4HCl_{(aq)} \rightarrow MnCl_{2(aq)} + Cl_2(g) + 2H_2O_{(l)}$

   (a) What does the 4 mean in 4HCl? ....

   (b) What does the 2 mean in MnO_2? ....

   (c) What do the symbols (s), (aq), (g) and (l) mean?

       (s) ....

       (aq) ....

(g) ....
(l) ....

Essay Section

1. Explain why equations must be balanced. Write a balanced equation for the reaction between copper and oxygen to form copper II oxide.
   For the above reaction write:-
   (i) a word equation
   (ii) a chemical equation
   (iii) a balanced equation explaining how you reached your final result. ....
   ....

2. Balance the following equations, briefly explaining your answer.
   (a) $N_2 + H_2 \rightarrow NH_3$
   (b) $SO_2 + O_2 \rightarrow SO_3$
   (c) $Al + Cl_2 \rightarrow AlCl_3$
   (d) $Mg + Cl_2 \rightarrow MgCl_2$
   (e) $NaNO_3 \rightarrow NaNO_2 + O_2$

4 — Oxygen

Multiple Choice Section

Questions 1-5 concern the following

    A — Catalyst
    B — Oxide
    C — Acidic
    D — Basic
    E — Amphoteric

Choose the correct answer from A to E

1. The oxides of non-metals are usually

2. Zinc oxide has acidic and basic properties because it is

3. When an element combines with oxygen it forms

4. Manganese IV oxide alters the speed of some reactions, in these reactions it is

5. The oxides of most metals are

Questions 6-10

For each of the following questions ONE or MORE of the answers are correct. Decide which of the responses is (are) correct. Then choose.

A	B	C	D	E
1,2,3 correct	1,2 only	2,3 only	1 only	3 only

6. Acidic oxides include
   1. carbon dioxide
   2. sulphur dioxide
   3. phosphorus V oxide

7. Catalysts
   1. alter the speed of a reaction
   2. are used up during the reaction
   3. change in mass

8. Elements whose oxides turn litmus blue include
   1. calcium
   2. sodium
   3. carbon

9. Which of the following processes use oxygen?
   1. mixing salt and sand
   2. breathing
   3. burning

16

10. Oxygen is NOT
    1. colourless
    2. odourless
    3. poisonous

    □

Questions 11-15

In the following questions you must decide if they are TRUE or FALSE. If true put T, if false put F.

11. A biological catalyst is called an enzyme. □

12. Water contains dissolved oxygen. □

13. Carbon dioxide turns damp red litmus blue. □

14. The oxygen molecule is diatomic. □

15. Magnesium burns very brightly in oxygen. □

Structured Section

1. This apparatus may be used in the preparation of oxygen.

(a) Name the chemicals A and B in the diagram

   A ....

   B ....

(b) Name the pieces of apparatus C, D, and E

   C ....

   D ....

E ....
(c) Give two properties of oxygen
   (i) ....
   (ii) ....
(d) How would you test the gas to show that it was oxygen?
....

2. Complete the following table.

Element	Name of oxide	Formula of oxide	Acidic, basic or neutral
Carbon			acidic
Hydrogen		H_2O	
Magnesium	Magnesium oxide		

(b) (i) Give the name of an oxide which reacts with water....................

(ii) Write an equation to show the reaction of this oxide with water..........
....

(c) Give the name of a metal which forms an oxide with both basic and acidic properties..

Essay Section

1. Oxygen is prepared in the laboratory by decomposing hydrogen peroxide, a catalyst is also used.
   (a) Explain what a catalyst is, and name the catalyst used in this preparation.
   (b) How is the gas collected?
   (c) Is oxygen soluble in water? Give a reason for your answer.
   (d) Give two main uses of oxygen.

2. Describe an experiment you have seen in which acidic and basic oxides were made. Your account should include a diagram, method and any observations made.

Practical Assessments

1. **Aim:** To find out if tap-water contains dissolved oxygen.

Apparatus: A round-bottomed flask, Bunsen burner, tripod, wire, gauze, safety mat, delivery tube, burette, trough, clamps and stands.

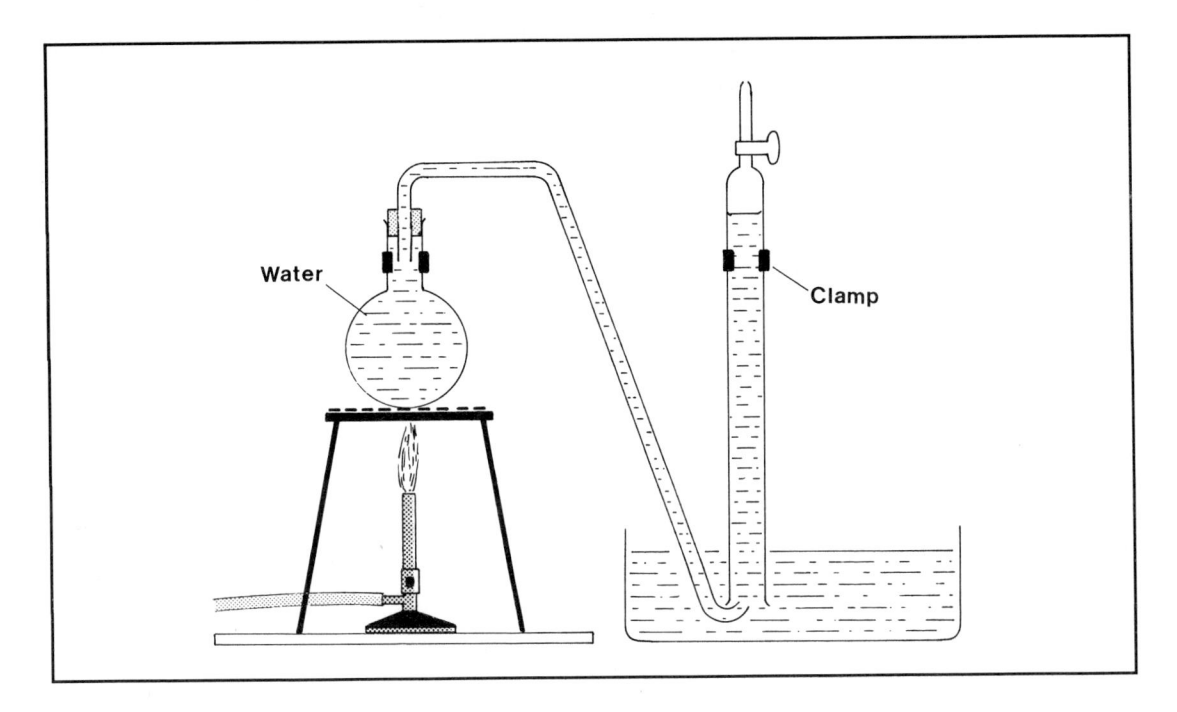

Method

1. Fill the flask and the delivery tube completely with water.
2. Fill a burette with water and up turn over the end of the delivery tube in a trough of water as shown in the diagram.
3. Heat the flask until the water boils.
4. Continue heating until there is no further change.
5. Record any changes.

2. **Aim:** To test the pH of solutions of the products of elements heated in air.
   Apparatus: Safety mat, safety glasses, tongs, crucible lid or tin, gas jars, deflagrating spoon, beakers, tripod, pipe-clay triangle.

Method

Part 1: Your teacher will demonstrate the heating of certain elements e.g. (phosphorus, sulphur, carbon, sodium). Watch closely what happens and record your observations in a table as shown at the end.

Part 2: You are provided with calcium, magnesium, copper, zinc and lead.

Calcium:

1. Using tongs hold a piece of calcium in a Bunsen flame.
2. When the reaction is complete, add the products of the reaction to a small amount of distilled water in a test-tube.

3. Add a drop of universal indicator or use indicator paper.

4. Record your results.

Magnesium: The same as calcium but DO NOT look at the flame. Copper, zinc and lead:

1. Put a sample of the metal on a crucible lid or tin lid on a pipe-lay triangle.

2. Heat the metal.

3. Add the product to a small amount of distilled water in a test-tube.

4. Test with universal indicator.

Results:

Element	Effect of heating	pH of solution

5 — Acids and Bases

Multiple Choice Section

Questions 1-5 concern the following pH values.

    A — 12
    B — 4
    C — 1
    D — 7
    E — 8

Choose from A to E the correct answer.

1. Which solution could be lemon juice? ☐

2. Which solution is the most acidic? ☐

3. Which solution is the most alkaline? ☐

4. Which solution is neutral? ☐

5. Which solution is distilled water? ☐

Questions 6-10

For each of the following questions ONE or MORE answers are correct. Decide which of the responses are correct. Then choose.

A	B	C	D	E
1,2,3 correct	1,2 only	2,3 only	1 only	3 only

6. Acids
   1. have a sweet taste
   2. turn litmus red
   3. have a low pH   ☐

7. A strong acid
   1. completely dissociates in water
   2. has a pH of 1 or 2
   3. often burns the skin   ☐

8. In a neutralisation
   1. a salt is formed
   2. the final solution has pH7
   3. water is formed   ☐

9. Sodium hydroxide is
   1. an alkali
   2. a base
   3. insoluble   ☐

10. An indicator
   1. is always colourless
   2. is a proton
   3. changes colour in acidic or basic solution ☐

Questions 11-15

In the following questions you must decide if they are TRUE or FALSE. If true put T, if false put F.

11. A titration is the controlled addition of an acid to an alkali. ☐

12. Vinegar is a strong acid. ☐

13. The pH scale is used to measure acidity and alkalinity. ☐

14. pH 12 is the value for a strong acid. ☐

15. Magnesium hydroxide is used to treat stomach upsets. ☐

Structured Section

1. In a titration to prepare sodium chloride Sodium hydroxide is reacted with hydrochloric acid using the following apparatus.

(a) Name the Apparatus A, B and C

   A ....
   B ....
   C ....

(b) Name an indicator you could use.
....
(c) What colour does the indicator change to?
....
(d) Why is C placed on a white tile?
....

2. Three solutions A, B and C have the pH values 3, 11, 7, respectively.
   (a) (i) Which is the most acidic? ....

   (ii) Which is the most alkaline? ....

   (b) Name substances which would have the same pH as solutions B and C

   B ....

   C ....

   (c) If an equal amount of solution A was mixed with an equal amount of solution B what would you expect the final pH to be? ....

   Explain your answer. ....
   ....

Essay Section

1. Describe how you would:
   (a) use a pipette to measure out exactly 25 cm^3 of a liquid.
   (b) use a burette to measure out exactly 25 cm^3 of a liquid.

2. Name the three strong acids used in the laboratory. What pH value would you expect these acids to have?
   Starting from sodium hydroxide describe how you would make sodium chloride crystals.

Practical Assessment

Aim: To make an indicator from flower petals or berries.

Apparatus: Mortar and pestle, rack of test-tubes, petals from red or blue flowers, or blackberries or elderberries, sodium hydrogen carbonate, ethanol, lemon juice.

Method

1. Grind the petals or berries with ethanol in a mortar.

2. When the liquid is a deep colour, pour it into a test-tube.

3. Put a few drops of coloured liquid into a second test-tube then add about 1 cm of lemon juice. Note the colour.

4. Add a small amount of sodium hydrogen carbonate, stir and note any colour change.

5. Continue adding sodium hydrogen carbonate until no further colour change takes place. Record the final colour.

6 — Preparation of Salts

Multiple Choice Section

Questions 1-5
The following questions are follwed by five possible answers A, B, C, D and E. Choose the BEST answer.

1. When an acid reacts with a carbonate
   - A — hydrogen is given off
   - B — carbon dioxide is given off
   - C — oxygen is given off
   - D — no gas is formed
   - E — sulphur dioxide is given off

2. When a salt is formed from a metal and an acid, which metal would NOT be used.
   - A — copper
   - B — magnesium
   - C — iron
   - D — zinc
   - E — tin

3. Salts are
   - A — covalent compounds
   - B — gases
   - C — ionic compounds
   - D — liquids
   - E — waxes

4. To make sure all the acid reacts in a salt preparation
   - A — the reaction is stirred
   - B — the reaction is heated
   - C — the mixture is filtered
   - D — excess solid is used
   - E — the filtrate is heated

5. An insoluble salt forms as
   - A — a precipitate
   - B — a filtrate
   - C — a solution
   - D — crystals
   - E — a gas

Questions 6-10

For each of the following questions ONE or MORE of the answers are correct. Decide which of the responses is (are) correct. Then choose.

A	B	C	D	E
1,2,3 correct	1,2 only	2,3 only	2 only	3 only

6. Salts can be
   1. soluble
   2. insoluble ☐
   3. covalent

7. Salts can be made from
   1. copper and hydrochloric acid
   2. copper carbonate and hydrochloric acid ☐
   3. copper II oxide and hydrochloric acid

8. All salts
   1. are solids
   2. are soluble ☐
   3. have low melting points

9. An acid will react with
   1. a base
   2. a carbonate ☐
   3. an alkali

10. The reaction between a metal and an acid is finished when
    1. it starts to bubble
    2. all the metal has disappeared ☐
    3. the bubbles stop

Questions 11-15

In the following questions you must decide if they are TRUE or FALSE. If true put T, if false put F.

11. Salts are always ionic. ☐

12. The filtrate is heated to concentrate the solution. ☐

13. Precipitates are always white. ☐

14. Excess solid is removed by evaporation. ☐

15. Slow cooling produces large crystals. ☐

Structured Section

1. (a) Zinc was added in excess to dilute hydrochloric acid. A gas and another chemical was formed.
      (i) What was the gas? ....
      (ii) What test could you use to identify the gas? ....
      (iii) Name the other substance formed ....

   (b) Calcium carbonate was added to dilute hydrochloric acid. A gas A and two other chemicals B and C were formed.
      (i) Name the gas A ....
      (ii) How would you test for gas A? ....

(iii) Name the substances B and C
   B ....
   C ....

2. Copper II sulphate (CuSO₄) is formed when excess copper II oxide CuO is added to dilute sulphuric acid and the mixture warmed. The copper II oxide is then removed.
   (a) Write a balanced equation for this reaction. ....
   (b) Why is excess copper II oxide used? ....
   (c) Why is the mixture heated? ....
   (d) How could the excess copper II oxide be removed? ....
   (e) Describe how large crystals of copper II sulphate could be obtained ....
   ....

Essay Section

1. Sodium sulphate can be prepared by gradually adding dilute sulphuric acid to sodium hydroxide solution and finding the neutral point using an indicator.
   (i) Draw the apparatus you would use.
   (ii) Name an indicator and state its colour change.
   (iii) Explain how crystals of sodium sulphate could be obtained.

2. Lead II iodide is an insoluble salt and can be prepared from lead II nitrate solution and potassium iodide solution. Describe how you would make this salt and obtain a dry sample of lead II iodide.

Practical Assessment

Aim: To prepare magnesium sulphate from magnesium and dilute sulphuric acid.

Apparatus: Beaker, glass stirring rod, filter paper, filter funnel, safety mat, tripod, wire gauze, Bunsen burner, evaporating dish, magnesium ribbon, dilute sulphuric acid.

Method

1. Measure out 30 cm³ of dilute sulphuric acid into a beaker.
2. Cut about 60 cm of magnesium ribbon into small pieces.
3. Add the magnesium to the dilute sulphuric acid.
4. When you think the reaction is finished, filter the mixture.
5. How will you obtain crystals of magnesium sulphate from the filtrate? When you think you know how, tell your teacher and if you are correct continue with the experiment.

Complete Science Work-Book 2

7 — Carbon and Carbon Dioxide

Multiple Choice Section

Questions 1-5 concern the following
    A — Diamond
    B — Allotrope
    C — Graphite
    D — Element
    E — Non-metallic.

Choose from A to E the correct answer.

1. Diamond and graphite are both forms of carbon. They are ☐

2. Carbon burns to form an acidic oxide. This is because it is ☐

3. The hardest naturally occurring substance ☐

4. This form of carbon conducts electricity ☐

5. Carbon is composed of only one type of atoms because it is ☐

Questions 6-10

For each of the questions ONE or MORE of the answers are correct. Decide which of the responses is (are) correct. Then choose.

A	B	C	D	E
1,2,3 correct	1,2 only	2,3 only	1 only	3 only

6. Carbon dioxide
   1. is a heavy gas
   2. does not support combustion
   3. is poisonous ☐

7. Diamond and graphite have the same
   1. structure
   2. appearance
   3. atoms ☐

8. Carbon and its compounds are found in
   1. plastics
   2. food
   3. plants ☐

9. Effervescence means
   1. loss of water of crystallisation
   2. bubbles of gas
   3. plants   ☐

10. Diamond and graphite are known as
    1. allotropes
    2. isotopes
    3. molecules   ☐

Questions 11-15

In the following questions you must decide if they are true or false. If true put T, if false F.

11. Carbon dioxide gas can be changed to a solid.   ☐

12. A lighted splint will burn in carbon dioxide.   ☐

13. Carbon dioxide forms carbonic acid in water.   ☐

14. Carbon dioxide is taken in by plants during photosynthesis.   ☐

15. Carbonic acid is a strong acid.   ☐

Structured Section

1. (a) When a gas jar containing carbon dioxide is held over a lighted candle in a beaker, the candle goes out.

What TWO properties of carbon dioxide does this show?
   (i) ....

(ii)...
....

(b) If a piece of burning magnesium is lowered into a gas jar of carbon dioxide, black specks form on the sides of the gas jar and a white ash falls to the bottom.
   (i) What are the black specks? ....
....
   (ii) What is the white ash? ....

2. (a) What are the two crystalline forms of carbon?
   (i) ....
   (ii) ....

(b) One of the forms is shown below. Complete the structure by adding more
   • carbon atoms and drawing the bonds. Also name this form.

Name ....

(c) Give TWO properties of the form shown in (b)
   (i) ....
   (ii) ....

Essay Section

1. Carbon dioxide can be prepared by reacting dilute hydrochloric acid with calcium carbonate ($CaCO_3$).
   (a) Draw the apparatus you would use.
   (b) What chemical could you use to dry the gas?
   (c) Write an equation for the reaction.

(d) How would you show that the gas was carbon dioxide? ....
....

2. Draw the two crystalline forms of carbon. What term is used to describe these forms? Give one use of each form and explain how the structure explains this use.

Practical Assessment

Aim: To study the reactions of carbon dioxide.

Apparatus: Boiling tube, delivery tube, small marble chips, dilute hydrochloric acid, test-tube, limewater, litmus solution and splints.

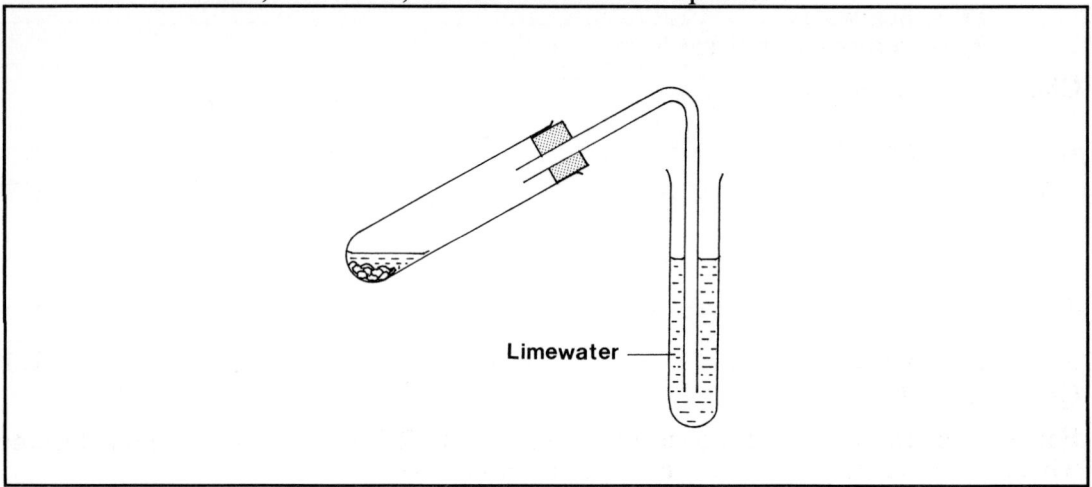

Limewater

Method

1. Put two spatula loads of marble chips in the boiling tube.
2. Put limewater into a test-tube.
3. Have stopper and delivery tube ready with the end of the delivery tube in the limewater.
4. Add dilute hydrochloric acid to the marble chips and QUICKLY connect the delivery tube. Record your results.
5. When the reaction has finished replace the test-tube containing the limewater by one containing litmus solution.
6. It may be necessary during the experiment to add more dilute hydrochloric acid.
7. Remove the delivery tube from the test-tube and hold a lighted splint near the end of the delivery tube. Record your results.

8 — Hardness of Water

Multiple Choice Section

Questions 1-5 concern the following

    A — Rain water
    B — Hard water
    C — Temporary hardness
    D — Sea water
    E — Permanent hardness

Choose the correct answer.

1. This can be removed by boiling ☐

2. This provides a good source of calcium in our diet ☐

3. One of the purest forms of water ☐

4. Contains many dissolved salts ☐

5. Contains calcium sulphate ☐

Questions 6-10

For each of the following questions ONE or MORE answers are correct. Decide which of the responses is (are) correct. Then choose

A	B	C	D	E
1,2,3	1,2	2,3	1	3
correct	only	only	only	only

6. Temporary hardness causes
   1. the formation of stalactites
   2. the formation of scale
   3. the formation of 'fur' ☐

7. All hardness can be removed by
   1. distillation
   2. ion exchange
   3. boiling ☐

8. Permanently hard water contains
   1. calcium sulphate
   2. calcium hydrogen carbonate
   3. sodium carbonate ☐

9. Rain water contains
   1. hydrogen
   2. carbon dioxide
   3. oxygen ☐

10. Disadvantages of hard water include
    1. wastes soap
    2. produces scum
    3. brewing beer ☐

Questions 11-15

In the following questions you must decide if they are TRUE or FALSE. If true put T, if false F.

11. Hard water easily forms a lather with soap. ☐

12. Soap reacts with Ca^{2+} ions in hard water. ☐

13. Limestone is a form of calcium carbonate. ☐

14. Hard water has a nicer taste. ☐

15. Deionised water is hard water. ☐

Structured Section

1. (a) Calcium carbonate occurs naturally as marble. Name TWO other forms of calcium carbonate which occur naturally.
   (i) ....
   (ii) ....

   (b)

   MARBLE
   CALCIUM CARBONATE —Heat→ SOLID X
   SOLID Y ←Add water— SOLID X

   (i) Give the chemical name of X ....

(ii) Give the chemical name of Y ....
(iii) Give the common name of Y ....
2. Samples of water were taken from three towns A, B, and C. The volume of soap solution needed to produce a lather before and after boiling was recorded.

	Volume of Soap before boiling	Volume of soap after boiling
A	10 cm^3	10 cm^3
B	10 cm^3	1 cm^3
C	10 cm^3	5 cm^3

(a) (i) Which town had permanently hard water? ....
(ii) Explain your answer ....
....

(b) (i) Which town had a mixture of permanent and temporary hardness? ....
....
(ii) Explain your answer ....
....

(c) Over what type of rocks do you think town B's water passed?
....

Essay Section

1. Rain water can go through many changes before it reaches the sea. Describe some of these changes and explain how they occur.
2. Explain what you understand by the term 'hard water'.
   Describe what you would see when soap is first added to hard water. Why does a lather eventually form?
   Name two chemicals which cause hardness in water and describe how one of them gets into the water.

Practical Assessment

Aim: To prepare a soap.

Apparatus: Test-tubes, beaker, tripod, wire, gauze, Bunsen burner, safety mat, filter funnel, filter paper, glass stirring rod, castor oil or olive oil, salt (sodium chloride), distilled water, 4M sodium hydroxide, a small measuring cylinder and safety glasses.

Method

1. Measure 2 cm^3 of olive oil or castor oil into a beaker.
2. Measure 10 cm^3 of sodium hydroxide and add to the oil in the beaker.
3. Boil the beaker for about 5 minutes.
4. Stir the contents of the beaker.
5. Measure 10 cm^3 of distilled water and add to the contents of the beaker.

6. Add about 5 spatula measures of salt to the beaker.
7. Boil the mixture for about 2 minutes, continue stirring.
8. Filter the contents of the beaker.
9. Half fill one test-tube with distilled water and a second test-tube with tap water. Add a small piece of soap to each.
10. Shake the test-tubes and record your results.

Complete Science Work-Book 2

9 — Oxidation and Reduction

Multiple Choice Section

Questions 1-5 concern the following

    A — Oxidation
    B — Reduction
    C — Redox
    D — Oxidising agent
    E — Reducing agent

Choose the correct answer from A to E.

1. A reaction in which electrons are gained ☐

2. This chemical removes electrons from another substance ☐

3. A reaction in which oxidation and reduction occurs ☐

4. A reaction in which oxygen is gained ☐

5. A reaction in whch electrons are lost ☐

Questions 6-10

For each of the following questions ONE or MORE of the answers are correct. Decide which of the responses is (are) correct. Then choose

A	B	C	D	E
1,2,3	1,2	2,3	1	3
correct	only	only	only	only

6. In which of the following is the underlined substance being oxidised?
   1. $\underline{2Fe^{2+}} + Cl_2 \rightarrow 2Fe^{3+} + 2Cl^-$
   2. $\underline{ZnO} + C \rightarrow Zn + CO$
   3. $\underline{CuO} + H_2 \rightarrow Cu + H_2O$ ☐

7. In which of the following does the particle on the left hand side lose electrons?
   1. $Zn_{(s)} \rightarrow Zu^{2+}{}_{(aq)}$
   2. $2Cl^-{}_{(aq)} \rightarrow Cl_2$
   3. $Fe^{3+}{}_{(aq)} \rightarrow Fe^{2+}{}_{(aq)}$ ☐

8. When potassium permanganate (potassium manganate VII) acts as an oxidising agent it becomes
   1. purple
   2. red
   3. colourless   ☐

9. When a chlorine atom becomes a chloride ion
   1. the chlorine atom gains one electron
   2. the chlorine atom gains one negative charge ☐
   3. the chlorine atom is reduced

10. In reduction
    1. oxygen is gained
    2. oxygen may be lost ☐
    3. electrons are gained

Questions 11-15

In the following questions you must decide if they are TRUE or FALSE. If true put T, if false F.

11. The oxidising agent is always reduced. ☐

12. Oxidation and reduction always occur together. ☐

13. In the reaction Na → Na+ the sodium atom is reduced. ☐

14. Carbon is often used as an oxidising agent. ☐

15. Rusting is oxidation. ☐

Structured Section

1. (a) What is oxidation? ....
   ....
   (b) Name two things which occur during oxidation
       (i) ....
       (ii) ....
   (c) Name an oxidising agent and briefly describe what you would see when it reacts.
       Oxidising agent ....
       When it reacts ....
       ....

2. Describe in terms of oxidation and reduction the following reactions saying which is reduced and which is oxidised and give a reason.
   (i) $H_{2(g)} + Cl_{2(g)} \rightarrow 2HCl_{(g)}$
       Which is reduced? ....
       Reason ....
       ....
       Which is oxidised? ....
       Reason ....
       ....

(ii) $Zn_{(s)} + H_2SO_{4(aq)} \rightarrow ZnSO_{4(aq)} + H_{2(g)}$

Which is reduced? ....

Reason ....

....

Which is oxidised? ....

Reason ....

....

Essay Section

1. Describe in terms of oxidation and reduction the following equations.
   (i) $Cl_{2(g)} + 2KBr_{(aq)} \rightarrow 2KCl_{(aq)} + Br_{2(l)}$
   (ii) $2PbO_{(s)} + C_{(s)} \rightarrow 2Pb_{(s)} + CO_{2(g)}$
   (iii) $Zn_{(s)} + CuSO_{4(aq)} \rightarrow ZnSO_{4(aq)} + Cu_{(s)}$

2. Say whether the following reactions are oxidation or reduction and explain your answers.
   (a) $Fe^{2+}_{(aq)} \rightarrow Fe^{3+}_{(aq)} + e$
   (b) $Mg_{(s)} \rightarrow Mg^{2+}_{(aq)} + 2e$
   (c) $G^+_{(aq)} + e \rightarrow Ag_{(s)}$
   (d) $Cl_{2(g)} + 2e \rightarrow 2Cl^-_{(aq)}$

Practical Assessment

Aim: To investigate the oxidising and reducing reactions of various chemicals. (Some of these reactions will be demonstrated by the teacher but you must watch carefully and record any observations).

Apparatus: Test-tubes, manganese(IV) oxide, potassium manganate (VII), potassium dichromate, concentrated hydrochloric acid, potassium iodide solution, potassium bromide solution, iron II sulphate, indicator paper.

Method
(To be demonstrated by the teacher)
1. Put small amount of manganese (IV) oxide, potassium manganate (VII) and potassium dichromate into separate test-tubes.
2. Add a few drops of concentrated hydrochloric acid to each test-tube.
3. Test any gas evolved with damp indicator paper or damp litmus paper.
4. Record your observations.
   (To be carried out by the pupil)
5. Pour a small amount of potassium bromide solution, potassium iodide solution and iron II sulphate solution into separate test-tubes.
6. Add a few drops of chlorine water to each test-tube.
7. Record your observations.

For each reaction you must say which chemical was reduced and which was oxidised.

10 — The Activity Series

Multiple Choice Section

Questions 1-5 concern the following

    A — Sodium
    B — Copper
    C — Hydrogen
    D — Alkali metals
    E — Metals

Choose from A to E the correct answer.

1. This metal burns with a yellow flame

2. Group 1 elements are known as

3. These tend to lose electrons to form positive ions

4. This metal will not react with acids to give hydrogen

5. This metal tarnishes in air

Questions 6-10

For each of the following questions ONE or MORE of the answers are correct. Decide which of the responses is (are) correct. Then choose

A	B	C	D	E
1,2,3	1,2	2,3	1	3
correct	only	only	only	only

6. The following metals will burn with a coloured flame
   1. lithium
   2. sodium
   3. potassium

7. The Activity Series is a series of
   1. atoms
   2. gases
   3. metals

8. Which of the following will react with acids to give hydrogen?
   1. zinc
   2. iron
   3. silver

9. Which of the following will displace copper from copper II sulphate?
   1. copper
   2. zinc
   3. iron ☐

10. The metals at the top of the series are
    1. very reactive
    2. slightly reactive
    3. unreactive ☐

Questions 11-15

In the following questions you must decide if they are TRUE or FALSE. If true put T, if false F.

11. Alkali metals are stored under oil. ☐

12. Copper occurs in the free state in nature. ☐

13. Magnesium will react with steam. ☐

14. Potassium burns with a lilac flame. ☐

15. Sodium is used in street lighting. ☐

Structured Section

1. The question concerns the following metals:
   aluminium, copper, calcium, zinc, lithium, potassium
   (a) Arrange the metals in order of reactivity, the MOST reactive first.
   ....
   (b) From the list choose a metal
      (i) That melts when added to cold water ....
      (ii) That does not react wth water or steam ....
      (iii) Whose compounds colour a Bunsen flame
         Metal ....
         Colour of flame ....
   (c) Aluminium will remove oxygen from iron III oxide in the Thermit reaction. Write an equation for the reaction.
      Fe_2O_3 + ...

2. The table below summarises the effect of adding small pieces of metal to aqueous solutions of metal salts.

Metal Added

Solution	Copper	Iron	Manganese	Silver
Copper II sulphate	—	copper displaced	copper displaced	no change
iron II sulphate	no change		iron displaced	no change
manganese II sulphate	no change	no change	—	no change
silver nitrate	silver displace	silver displaced	silver displaced	

(a) (i) Arrange the four metals in order of reactivity, MOST reactive first.........
　　(ii) Give a reason for the order in (i) ....
　　..
(b) Describe TWO changes you would SEE when excess iron filings were added to copper II sulphate.
　　(i) ....
　　(ii) ....
(c) Write an equation for the reaction between iron and copper II sulphate.
　　..

Essay Section

1. (a) Explain why an iron nail file becomes covered with a pink layer if placed in copper II sulphate solution.
　(b) Explain why a piece of copper becomes coated with a pale grey layer when placed in silver nitrate solution.
　　Arrange copper, iron and silver in order of increasing reactivity.
2. When copper II oxide (a back powder) is heated with black carbon in a test-tube, the mixture glows and a red solid forms. Explain what has happened and write an equation for the reaction. Has oxidation or reduction occurred?

Practical Assessment

Aim: To find the position of carbon in the Activity Series.
Apparatus: Test-tubes, safety glasses, safety mat, test-tube holder, charcoal powder, copper II oxide, magnesium oxide and lead II oxide.

Method

1. Put equal amounts of charcoal powder and copper II oxide in a test-tube.
2. Heat the mixture until it glows.
3. Test for carbon diode by holding a glass rod with a drop of limewater on it near

the neck of the test-tube.
4. Cool the test-tube and record the appearance of the residue.
5. Repeat this experiment with magnesium oxide, then with lead II oxide.
6. Record your results and try to place carbon in a position with respect to copper, magnesium and lead.

11 — Sulphur and Sulphur Dioxide

Multiple Choice Section

Questions 1-5 concern the following

    A — Sulphur
    B — Sulphur dioxide
    C — Plastic sulphur
    D — Monoclinic sulphur
    E — Rhombic sulphur

Choose from A to E the correct answer.

1. It is found in large quantities in Texas ☐

2. Formed when boiling sulphur is added to cold water ☐

3. One of the main causes of acid rain ☐

4. Takes the form of long needle-shaped crystals ☐

5. This can be used as a food preservative ☐

Questions 6-10

For each of these questions ONE or MORE of the answers is correct. Decide which of the responses is (are) correct. Then choose.

A	B	C	D	E
1,2,3, correct	1,2 only	2,3 only	1 only	3 only

6. The crystalline allotropes of sulphur are
   1. plastic sulphur
   2. rhombic sulphur
   3. monoclinic sulphur ☐

7. Sulphur is used in the manufacture of
   1. sulphuric acid
   2. vulcanised rubber
   3. iron pyrities ☐

8. Sulphur dioxide is
   1. colourless
   2. heavy
   3. poisonous ☐

9. Sulphur dioxide is dried by
   1. concentrated sulphuric acid
   2. calcium oxide
   3. magnesium oxide ☐

10. Sulphur burns with a
    1. yellow flame
    2. white flame
    3. blue flame ☐

Questions 11-15

In the following questions you must decide if they are TRUE or FALSE. If true put T, if false F.

11. Sulphur dioxide is used in wine making. ☐

12. Plastic sulphur is a crystalline allotrope of sulphur. ☐

13. Sulphur dioxide will decolourise potassium manganate VII. ☐

14. Sulphur is a non-metal. ☐

15. Sulphur dioxide turns potassium dichromate green. ☐

Structured Section

1. (a) When sulphur is heated in air strong smelling gas is formed.
      (i) Name the gas formed ....
      (ii) Write an equation for its formation
         ....
      (iii) How would the gas affect damp litmus paper?
         ....
   (b) The gas reacts with water to form a new substance.
      (i) Name the new substance formed ....
      (ii) Write an equation to show the formation of the new substance.
         ....

2. Sulphur dioxide reacts with oxygen in the presence of a catalyst according to the following equation

$$2SO_2 + O_2 \rightleftharpoons 2SO_3$$

   (a) Explain what is meant by a catalyst.
      ....
      ....
      ....

   (b) What does the sign $\rightleftharpoons$ mean?

....
....
(c) Give TWO uses of sulphur dioxide
  (i) ....
  (ii) ....

Essay Section

1. Describe how DRY sulphur dioxide can be prepared in the laboratory. Your answer should include a diagram, method and equation.
What is the test for sulphur dioxide?

2. Starting from powdered sulphur describe how you could prepare the two crystalline allotropes of sulphur. In each case draw the crystal formed.

Practical Assessment

Aim: To study the effect of sulphur dioxide on natural materials.

Apparatus: Test-tubes, corks, test-tube rack, Bunsen burner, safety mat, safety glasses, wool, silk, straw, flower petals, sodium sulphite, dilute hydrochloric acid, delivery tube and test-tube holders.

Method

1. Put wool in one test-tube, straw in another, silk in another and a coloured flower petal in a fourth test-tube.

2. Put two spatula measures of sodium sulphite in a fifth test-tube.

3. Add dilute hydrochloric acid to the sodium sulphite to a depth of 2 cm.

4. Put a cork containing a delivery tube into the fifth test-tube.

5. Warm the test-tube containing the sodium sulphite and put the end of the delivery tube into a test-tube containing one of the materials.

6. After about 2 minutes cork the test-tube containing the materials and leave.

7. Repeat 5 and 6 with each of the materials.

12 — Ammonia

Multiple Choice Section

Question 1-5 concern the following

    A — Nitrogen
    B — Calcium oxide
    C — Hydrogen chloride
    D — Litmus
    E — Calcium hydroxide

Choose from A to E the correct answer.

1. Ammonia is one of the most important compounds of ☐

2. Ammonia forms dense white fumes with ☐

3. Slaked lime is ☐

4. Ammonia is dried using ☐

5. This is turned blue by ammonia ☐

Questions 6-10

For each of the questions ONE or MORE of the answers are correct. Decide which of the responses is (are) correct. Then choose

A	B	C	D	E
1,2,3 correct	1,2 only	2,3 only	1 only	3 only

6. Ammonia is
   1. lighter than air
   2. heavier than air
   3. as dense as air
      ☐

7. The main source of nitrogen for plants is
   1. the air
   2. manure
   3. fertilisers
      ☐

8. Ammonia can be dried using
   1. concentrated sulphuric acid
   2. calcium chloride
   3. calcium oxide
      ☐

9. Ammonia is the only common
   1. alkaline gas
   2. acidic gas
   3. neutral gas

   □

10. Ammonia is used in the manufacture of
    1. household cleaners
    2. nitric acid
    3. nitrogen

    □

Questions 11-15

In the following questions you must decide if they are TRUE or FALSE. If true put T, if false put F.

11. Ammonia is not very soluble in water. □

12. Ammonia has no smell. □

13. The formula for ammonia is NH_3. □

14. Ammonia will support combusion. □

15. Concentrated ammonia is often called 880 ammonia. □

Structured Section

1. Sometimes farmers spread chemicals on the ground to provide food for the plants.
   (a) (i) What is the general name given to these chemicals?
   ....
   (ii) Give the chemical name of one of these substances.
   ....
   (b) Ammonia is a very useful chemical
   (i) What TWO elements are present in ammonia?
   ....
   (ii) Draw the shape of an ammonia molecule

   (iii) What type of bonds are present in ammonia?
   ....

2.

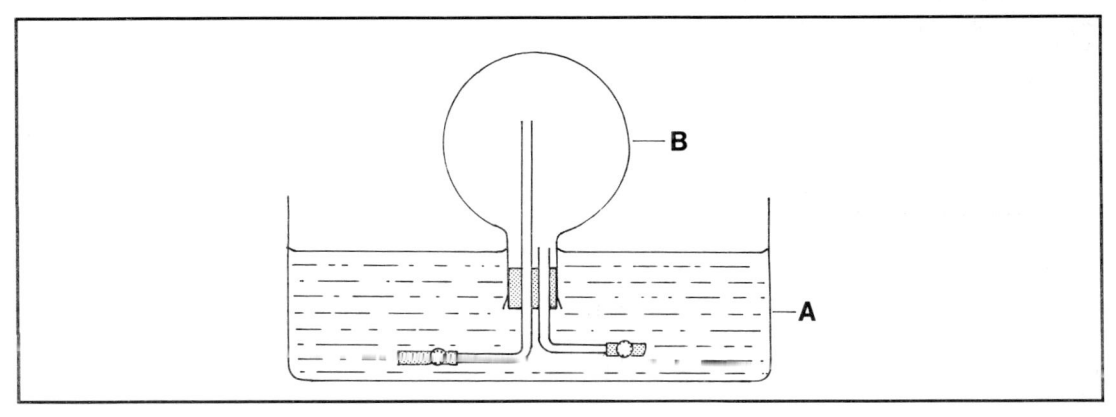

(a) (i) what is in trough A?
....
(ii) What is in flask B?
....

(b) What happens when the clip on the tube is opened?
....
....

(c) What does this experiment tell you about the solubility of ammonia?
....
....

Essay Section

1. Ammonia can be made in the laboratory by heating an ammonium compound with an alkali.
   (a) Name TWO chemicals you could use for this preparation.
   (b) Write an equation for the reaction between the chemicals you chose in (a).
   (c) Ammonia is a soluble, light gas, how do these properties effect the way in which you would collect gas?

2. Draw and label a diagram for the laboratory preparation of ammonia. Describe what would happen if:
   (i) Ammonia came in contact with damp red litmus.
   (ii) Ammonia came in contact with hydrogen chloride.
   Explain your answers and give an equation for (ii)

Practical Assessment

Aim: To find the effect of aqueous ammonia on aqueous solutions of certain metal salts.

Apparatus: Test-tubes, test-tube rack, ammonium hydroxide, solutions of copper II sulphate, iron II sulphate, iron III sulphate, lead II nitrate, and zinc sulphate.

Method

1. Put one of the metal salt solutions into a test-tube (about 2cm depth).
2. Add ammonium hydroxide DROP BY DROP until no further change occurs.
3. Record your results.
4. Repeat for each metal salt using a clean test-tube each time.

Results

Metal Salt	Drops of ammonium hydroxide	Excess ammonium hydroxide

13 — Chlorine

Multiple Choice Section

Questions 1-5 concern the following:

    A — Chlorine
    B — Astatine
    C — Fluorine
    D — Iodine
    E — Bromine

Choose from A to E the correct answer.

1. Is a dark red liquid. ☐
2. Used in swimming pools. ☐
3. One of its compounds is used in toothpaste. ☐
4. A dark grey solid. ☐
5. A very rare halogen. ☐

Questions 6-10

For each of the questions ONE or MORE of the answers are correct. Decide which of the responses is (are) correct. Then decide

A	B	C	D	E
1,2,3 correct	1,2 only	2,3 only	1 only	3 only

6. Chlorine is used in the manufacture of
   1. hydrochloric acid
   2. dry cleaning solvents
   3. anaesthetics
                                                                                     ☐

7. Chlorine is
   1. in Group 7 of the Periodic Table
   2. a non-metal
   3. a noble gas
                                                                                     ☐

8. Chlorine occurs
   1. in the atmosphere
   2. in the free state
   3. in rock salt
                                                                                     ☐

9. Chlorine should be prepared
   1. in a fume cupboard
   2. in a laboratory
   3. in a classroom

10. Chlorine is
    1. colourless
    2. heavy
    3. soluble in water

Questions 11-15

In the following questions you must decide if they are TRUE or FALSE. If true put T, if false put F.

11. Chlorine is a good oxidising agent.

12. Chlorine will not react with non-metals.

13. Chlorine reacts with metals to form salts.

14. Chlorine is a bleaching agent.

15. Chlorine is not poisonous.

Structured Section

1. When a mixture of concentrated hydrochloric acid and potassium manganate VII are heated together a green gas is given off.
   (a) Name the green gas ....
   (b) (i) What is the potassium manganate VII acting as?
       ....
       (ii) Write an equation to show what has happened to the chloride ions in hydrochloric acid.
       ....
   (c) How could the gas be dried? ....
       ....
   (d) How would the green gas affect damp litmus paper? ....
       ....

2. Describe what you would SEE when chlorine comes in contact with the following:
   (a) Potassium bromide solution ....
       ....
       Write the equation for the reaction ....
       ....
   (b) Potassium iodide solution ....

....
Write the equation for the reaction ....
....

(c) Red rose petals ....
....

Essay Section

1. Fluorine, chlorine and iodine belong to the same chemical family.
   (a) What is the name of this family? Also name another element which belongs to this family and describe its appearance.
   (b) Chlorine Atomic number 17 forms a diatomic molecule. Show how the bonding occurs between the chlorine atoms.
2. Describe how chlorine could be prepared in the laboratory. Draw a diagram and say where the preparation should be carried out. How could you prove that the gas was chlorine? Give TWO uses of chlorine.

Practical Assessment

Aim: To see how the halides react with silver nitrate solution.

Apparatus: Test-tubes, test-tube rack, filter paper, filter funnel, silver nitrate solution, potassium chloride, potassium bromide, potassium iodide, distilled water and nitric acid.

Method

1. Put a spatula measure of potassium chloride in a test-tube.
2. Add distilled water (about 2cm depth) and dissolve the potassium chloride.
3. Add a few drops of dilute nitric acid.
4. Add silver nitrate solution drop by drop until the reaction is complete. Record your results.
5. Filter the contents of the test-tube.
6. Remove the filter paper and leave to dry. Record your observations.
7. Repeat 1-6 for potassium bromide and potassium iodide using a clean test-tube each time.

14 — Thermochemistry

Multiple Choice Section

Questions 1-5 concern the following

    A — Endothermic
    B — Heat change
    C — Positive
    D — Exothermic
    E — Negative

Choose from A to E the correct answer.

1. When a fuel burns the reaction is. ☐

2. This type of reaction takes in heat. ☐

3. ΔH means. ☐

4. ΔH is positive when the reaction is. ☐

5. In an ethermic reaction H is. ☐

Questions 6-10

For each of the questions ONE or MORE of the answers are correct. Decide which of the responses is (are) correct. Then choose

A	B	C	D	E
1,2,3	1,2	2,3	1	3
correct	only	only	only	only

6. The heat of neutralisation is the heat produced when which of the following react?
    1. an acid
    2. a base
    3. water

    ☐

7. -55 Kjoules means
    1. you must heat the reaction
    2. the reaction is endothermic
    3. the reaction is exothermic

    ☐

8. A mole of substance is
    1. the mass used
    2. it's relative atomic mass in grams
    3. it's relative molecular mass in grams

    ☐

9. In an endothermic reaction
   1. the temperature falls
   2. heat is taken in
   3. Δ H is positive □

10. Heat change is measured in
    1. Kilojoules
    2. °C
    3. cm^3 □

Questions 11-15

In the following questions you must decide if they are TRUE or FALSE. If true put T, if false F.

11. Heat of solution is always exothermic. □

12. When coal burns the reaction is exothermic. □

13. For an exothermic reaction Δ H is positive. □

14. Δ H represents only the heat taken in. □

15. Heat of neutralisation is always exothermic. □

Structured Section

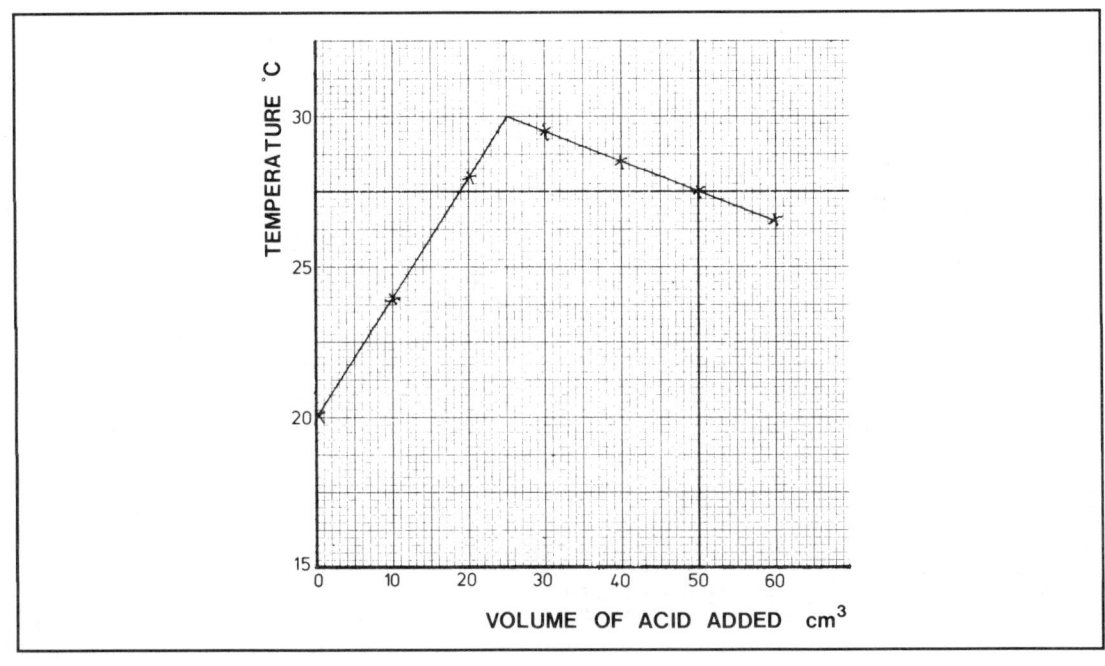

1. The graph shows the maximum temperatures recorded when the volumes of hydrochloric acid shown are added to 25 cm^3 of 1M sodium hydroxide solution in a plastic cup. The solutions were both at room temperature before mixing.

Use the graph to answer the following:—
(a) What was the room temperature? ....
(b) (i) Was the reaction between hydrochloric acid and sodium hydroxide endothermic or exothermic? ....
   (ii) Explain your answer ....
   ....
(c) What was the highest temperature recorded? ....
(d) Write the equation for the reaction ....
....

2. (a) When silver nitrate solution reacts with a solution sodium chloride the energy change is $\Delta H = -65.7$ Kjoules per mole.
   (i) Explain what ΔH means ....
   ....
   (ii) Explain what -65.7 KJ means ....
   ....
   (iii) Explain what is meant by a mole ....
   ....

(b) In the manufacture of ammonia the equation is
$$N_{2(g)} + 3H_{2(g)} \rightleftharpoons 2NH_{3(g)} \qquad \Delta H = -93 \text{ KJ per mole}$$
   (i) What do you understand by the sign $\rightleftharpoons$?
   ....
   (ii) Is the reaction given endothermic or exothermic?
   ....

Essay Section

1. Give one example of a solid fuel, a liquid fuel and a gaseous fuel.
   (i) What type of reaction occurs when these fuels are burned. Explain your answer.

   (ii) Would ΔH be positive or negative.

2. When 1 mole of solid ammonium nitrate dissolves in 1dm³ of water the heat change was $+25$ KJ per mole. Explain the meaning of this.
   Describe an experiment you could carry out to find the heat of solution of ammonium nitrate.

Practical Assessment

Aim: To find the effect of water on different solutes.

Apparatus: Insulated cup and lid, thermometer, distilled water, ammonium chloride, anhydrous and hydrated copper II sulphate and ammonium nitrate.

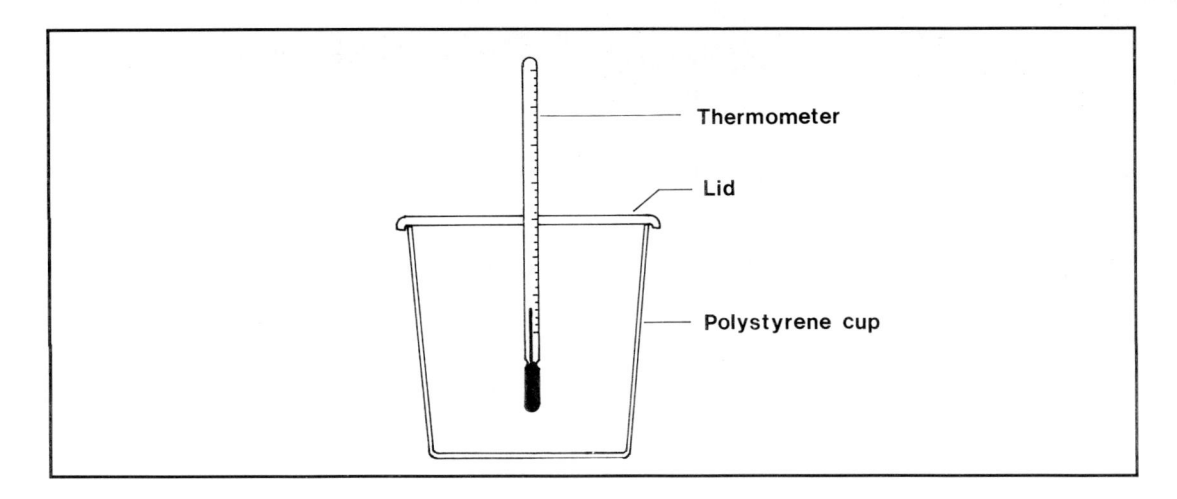

Method

1. Half fill the plastic cup with distilled water.
2. Replace the lid and measure the temperature of the water. Record the temperature . . . . °C.
3. Remove the lid and quickly add about three spatula measures of ammonium chloride.
4. Replace the lid and stir the contents with the thermometer.
5. Record the highest temperature reached. . . . . °C.
6. Wash the insulated cup and then repeat 1-5 for the other substances.

15 — Chemical Calculations

Relative Molecular Mass R.M.M.

Relative molecular mass is found by adding together the relative atomic masses of all the atoms in the molecule.

A list of relative atomic masses will be found at the end.

1. Find the relative molecular mass of each of the following:

   CO_2 $CaSO_4$
   NO_2 $Mg(OH)_2$
   SO_2 Al_2O_3
   P_2O_5 $FeSo_4$
   Na_2CO_3 K_2SO_4
   HCl $FeCl_3$
   KOH $NaHCO_3$
   $AlCl_3$ $Al(OH)_3$
   H_2SO_4 H_2CO_3
   $CaCl_2$ $NaNO_3$

Percentage Composition

Remember:

$$\frac{(\text{R.A.M. of the element} \times \text{number of atoms})}{\text{R.M.M.}} \times 100 = \% \text{ of the element}$$

2. Calculate:
   (i) The R.M.M. of the compound
   (ii) The percentage by mass of each element in the compound.

   (a) Carbon and oxygen in CO.
   (b) Sodium and chlorine in NaCl.
   (c) Carbon and hydrogen in C_2H_2
   (d) Potassium, ogen and hydrogen in KOH.
   (e) Aluminium and oxygen in Al_2O_3
   (f) Nitrogen and hydrogen in NH_3.

Relative Atomic Mass R.A.M.

Aluminium	Al	27
Calcium	Ca	40
Carbon	C	12
Chlorine	Cl	35.5
Hydrogen	H	1
Iron	Fe	56
Magnesium	Mg	24

Complete Science Work-Book 2

Nitrogen	N	14
Oxygen	O	16
Phosphorus	P	31
Potassium	K	39
Sodium	Na	23
Sulphur	S	32

16 — Food Types

Multiple Choice Section

Questions 1-5

Each question is followed by five possible answers A, B, C, D and E. Choose the BEST answer.

1. What is the name of the storage carbodydrate found in most plants?
   - A — Sugar
   - B — Glycogen
   - C — Starch
   - D — Glucose
   - E — Sucrose

2. All fats consist of
   - A — Fatty acids only
   - B — Glycerol only
   - C — Polyunsaturated fats
   - D — Grease
   - E — Fatty acids and glycerol

3. Iodine solution can be used to test for starch. If starch is present the iodine turns to?
   - A — Blue-black
   - B — Red
   - C — Yellow
   - D — Green
   - E — Orange

4. To test a food substance for the presence of sugars you would add which of the following substances?
   - A — Copper sulphate solution
   - B — Benedict's solution
   - C — Iodine solution
   - D — Sodium hydroxide
   - E — Sudan III

5. The main function of protein is for?
   - A — Energy
   - B — Storage
   - C — Protection
   - D — Growth and repair of cells
   - E — Insulation

Questions 6-10 concern the following
   - A — Vitamin A
   - B — Vitamin C
   - C — Vitamin K
   - D — Vitamin D
   - E — Vitamin B

Science Work-Book 2

Choose from A to E the BEST answer.

6. Which vitamin is found in carrots? ☐
7. Scurvy is caused by deficiency of this vitamin? ☐
8. Which vitamin is necessary for healthy bones and teeth? ☐
9. Which vitamin is needed for the clotting of the blood? ☐
10. Yeast is rich in which vitamin? ☐

Questions 11-15

In the following questions you must decide if they are TRUE or FALSE. If true put T, if false put F.

11. Carbohydrates consist of carbon, hydrogen, oxygen and nitrogen. ☐
12. An oil is a liquid fat. ☐
13. Lean meat and fish are good sources of protein. ☐
14. A balanced diet consists of carbohydrates, fats, proteins, vitamins, minerals, fibre and water. ☐
15. Minerals are necessary in large quantities. ☐

Structured Section

1. a) Describe the structure of a carbohydrate............................
   ....
   ....
   ....
   ....

   b) Give 3 examples of carbohydrate-rich foods ....
   ....
   ....
   ....

   c) What are the functions of fats?..................................
   ....
   ....
   ....

   d) What are the main functions of protein in the body?...............
   ....
   ....
   ....

e) Why is fibre a necessary part of our diet? ....
....
....
....

2. a) What is meant by a balanced diet? ....
....
....
....

b) Why are vitamins a necessary part of our diet? ....
....
....
....

c) Name 3 vitamins. State which disease is associated with a deficiency of each one and indicate which foods should be eaten to prevent the deficiency disease occurring.
   (i) Vitamin ....
   ....
   (ii) Vitamin ....
   ....
   (iii) Vitamin ....
   ....

Essay Section

1. State what is meant by a balanced diet. Describe each of the food types that make up a balanced diet.
2. Make a list of all the food and drink that you have consumed in the last 24 hours. Decide whether your diet is a good balanced diet and state how you could improve it, if necessary

Practical Assessment

Aim: To test various foods for the presence of starch.

Apparatus: Test-tubes, knife, tile, iodine solution.
A selection of foods e.g. potato, rice, bread, apple, orange, fish, meat, eggs, cheese, etc.

Method

1. Use a knife and tile and chop up a piece of potato.
2. Drop some iodine onto the potato.
3. Look for the appearance of the blue-black colour which tells you that starch is

present.
4. Repeat steps 1-3 using each of the other available foods. Use a test-tube if the food is a liquid.
5. Complete the following result table.

Results

Food	Colour of iodine after test	Starch present Yes/No
Potato egg rice bread apple orange fish meat cheese		

Science Work-Book 2

17 — Animal Nutrition

Multiple Choice Section

Questions 1-5 concern the following:
- A — ingestion
- B — digestion
- C — absorption
- D — assimilation
- E — egestion

Choose from A to E the BEST answer.

1. Large particles of food are broken down into smaller particles by this process. ☐

2. This is the removal of unused food from the body. ☐

3. The taking in of food through our mouths. ☐

4. Food passes through the gut wall and into the bloodstream. ☐

5. Small particles of food are built up into larger particles within the body. ☐

Questions 6-10 concern the following diagram.

Choose from A to E the BEST answer.

6. The main function of this region is to re-absorb water. ☐

7. Food remains here for about 3 hours. ☐

8. This is called the appendix. ☐

9. Digested food is absorbed in this region. ☐

10. This is where unused food leaves the body. ☐

Questions 11-15

In the following questions you must decide if they are TRUE or FALSE. If true put T, if false put F.

11. The human alimentary canal is approximately 10 metres long. ☐

12. An enzyme is a biological catalyst. ☐

13. The stomach has an alkaline pH. ☐

14. Food moves down the oesophagus by peristalsis. ☐

15. Amylase breaks down fats into fatty acids and glycerol. ☐

Structured Section

1. a) Why do we need to eat food? ....
   ....
   ....
   ....

   b) What is digestion? ....
   ....
   ....
   ....

   c) Name 3 digestive enzymes. State the chemical changes that they bring about.
   1. Enzyme ....
      Chemical change ....
   2. Enzyme ....
      Chemical change ....
   3. Enzyme ....
      Chemical change ....

2. Describe the functions of the following parts of the alimentary canal.
   a) Mouth ....

7. The hardest substance in the body. ☐

8. The 'living' part of the tooth. ☐

9. The root is embedded in this. ☐

10. Most of the tooth consists of this. ☐

Questions 11-15

In the following question you must decide if they are TRUE or FALSE. If true put T, if false put F.

11. The crown is the visible part of the tooth. ☐

12. Man has 32 teeth. ☐

13. The correct name for tooth decay is CARIES. ☐

14. Incisors are used for grinding food. ☐

15. Dental decay is caused by sugar. ☐

Structured Section

1. a) Draw a diagram of a tooth. Label the following parts: Enamel, dentine, pulp cavity, root, cement, periodontal fibres.

b) Describe the function of each.

Enamel ....
....
Dentine ....
....
Pulp cavity ....
....
Periodontal fibres ....
....

Cement..
....

2. a) What is the correct name for tooth decay? ....
....

b) What is plaque? ....
....

c) How is plaque involved in tooth decay? ....
....

d) Why is fluoride sometimes added to toothpaste? ....
....

e) What are disclosing tablets used for? ....
....

Essay Section

1. Describe a good plan of dental hygiene. Include in your answer the following:— dental visits, brushing of teeth, good eating habits.
2. Man is an omnivore. Describe how you would expect his teeth to differ from the teeth of a NAMED carnivore and a NAMED herbivore.
3. Describe the part played by plaque in tooth decay.

Practical Assessment

1. **Aim:** To show plaque on teeth.

    Apparatus: Disclosing tablets, toothbrush, toothpaste, mirror, stop-clock.

Method

1. Chew on a disclosing tablet. Use the mirror to look at the extent of staining on your teeth. Plaque is stained by the disclosing tablet.
2. Start the stop-clock and begin cleaning your teeth, using the toothbrush and toothpaste.
3. Note the time taken to remove all the stain, i.e. plaque, from your teeth.

Results

Time taken to remove plaque from teeth = ...

Answer the following questions:—

1. What are disclosing tablets used for? ....
2. How long should you spend, brushing your teeth to ensure that all plaque is removed. ....

Science Work-Book 2

19 — Plant Nutrition

Multiple Choice Section

Questions 1-5

Each of the following are followed by five possible answers A,B,C,D and E. Choose the BEST answer.

1. Which of the following factors is NOT necessary for photosynthesis to take place?
    - A — Carbon dioxide
    - B — Light
    - C — Chlorophyll
    - D — Water
    - E — Oxygen   ☐

2. Where does photosynthesis take place?
    - A — Leaves only
    - B — Stem only
    - C — Roots
    - D Green parts of a plant
    - E All over the plant   ☐

3. Which of the following is the correct equation for photosynthesis?

    A — $6CO_2 + 6H_2O \rightarrow C_6H_{12}O_6 + 6O_2$

    B — $CO_2 + H_2O \rightarrow C_6H_{12}O_6 + O_2$

    C — $6CO_2 + 6H_2O \xrightarrow[\text{chlorophyll}]{\text{light}} C_6H_{12}O_6 + 6O_2$

    D — $CO_2 + H_2O \xrightarrow[\text{chlorophyll}]{\text{light}} C_6H_{12}O_6 + O_2$

    E — $6CO_2 + 6H_2O \xrightarrow[\text{chlorophyll}]{\text{light}} C_6H_{12}O_6$ ☐

4. The following experiment is used to show:—

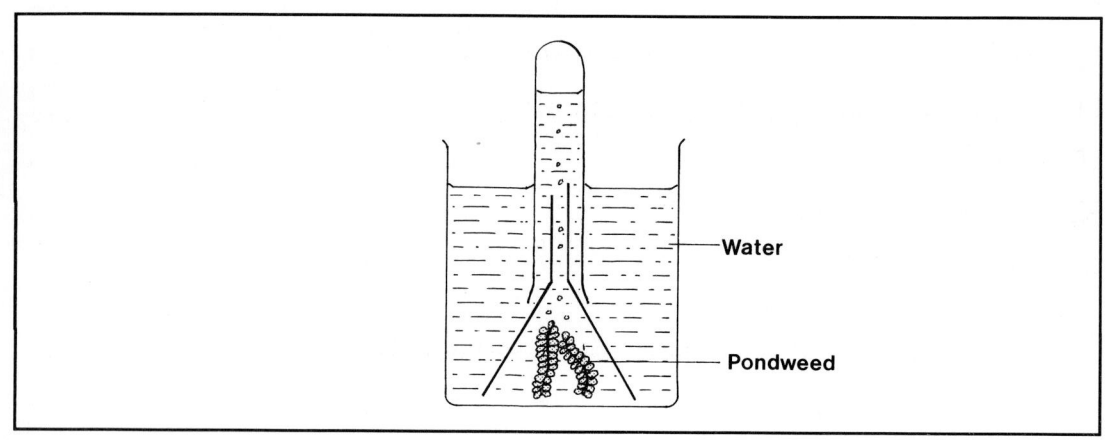

A — Oxygen is released during photosynthesis
B — Oxygen is required for photosynthesis
C — Chlorophyll is necessary for photosynthesis
D — Carbon dioxide is necessary for photosynthesis ☐
E — Carbon dioxide is produced during photosynthesis

5. How would you destarch a plant?
   A — Don't water a plant for 48 hours
   B — Leave a plant in the dark for 48 hours
   C — Leave a plant in the dark for 24 hours
   D — Cover a plant with a plastic bag for 24 hours ☐
   E — Cover a plant with a plastic bag for 48 hours

Questions 6-10 concern the following:—
   A — Magnesium
   B — Nitrogen
   C — Calcium
   D — Potassium
   E — Phosphorus

Choose the BEST answer from A to E

6. Which mineral has the symbol K. ☐

7. Which mineral acts as a cement to bind cells together. ☐

8. Which mineral is required for chlorophyll production. ☐

9. Which mineral is necessary for protein formation in plants. ☐

10. A deficiency of which mineral, results in slow growth with dull green leaves. ☐

Questions 11-15

In the following questions you must decide if they are TRUE or FALSE. If true put T if false put F.

11. Plants make their own food by photosynthesis. ☐

12. Chlorophyll is required for photosynthesis. ☐

13. Carbon dioxide is produced by photosynthesis. ☐

14. Minerals are necessary for the healthy growth of plants. ☐

15. A variegated leaf, is one which has white and green patches. ☐

Structured Section

1. a) What is photosynthesis? ....
   ....
   b) List the factors that are required for photosynthesis to take place ....
   ....

c) Where does photosynthesis take place? ....

d) What are stomata? Where would you find them? ....

2. a) Discuss the importance of the following minerals in plant growth.

   A Nitrogen ....

   B Phosphorus ....

   C Iron ....

   b) Give details of an experiment that shows the importance of minerals in plant growth ....

   c) Discuss why a farmer uses fertilisers ....

Essay Section

1. Give the equation for photosynthesis. Describe an experiment to show that carbon dioxide is necessary for photosynthesis.
2. Give details of an experiment to show that oxygen is released during photosynthesis. Explain what happens to the oxygen that is produced.

Practical Assessment

1. **Aim:** To show that chlorophyll is necessary for photosynthesis.

   Apparatus: Variegated leaves (e.g. Coleus or Tradescantia), beakers, tripod Bunsen-burner, alcohol, iodine solution, white tile, test-tubes.

Method

1. Draw a diagram of the variegated leaf, showing clearly any green areas.
2. Test the leaf for the presence of starch, using the following method.

   (i) Bring a water bath to the boil, and drop the leaf into it. This kills the leaf and stops any further enzyme activity. Turn off the Bunsen burner, as the next step involves alcohol, which is inflammable.

   (ii) Remove the leaf and allow it to cool.

   (iii) Warm a test-tube of alcohol by placing it in the water bath. Place the leaf in the warm alcohol and leave for about 10 minutes. The alcohol removes the green colour from the leaf, ensuring that any future colour change will be obvious.

(iv) Remove the leaf from the test-tube and dip it into the water bath to soften it, and place it on a white tile.

(v) Drop iodine solution onto the leaf and look for the appearance of blue-black areas, indicating the presence of starch.

3. Draw a second diagram of the leaf marking any blue-black areas, i.e. those areas that contain starch.
4. Compare this diagram with the diagram of the leaf at the start of the experiment. You should find these diagrams are identical, showing that chlorophyll is necessary for starch production i.e. photosynthesis.

Summary

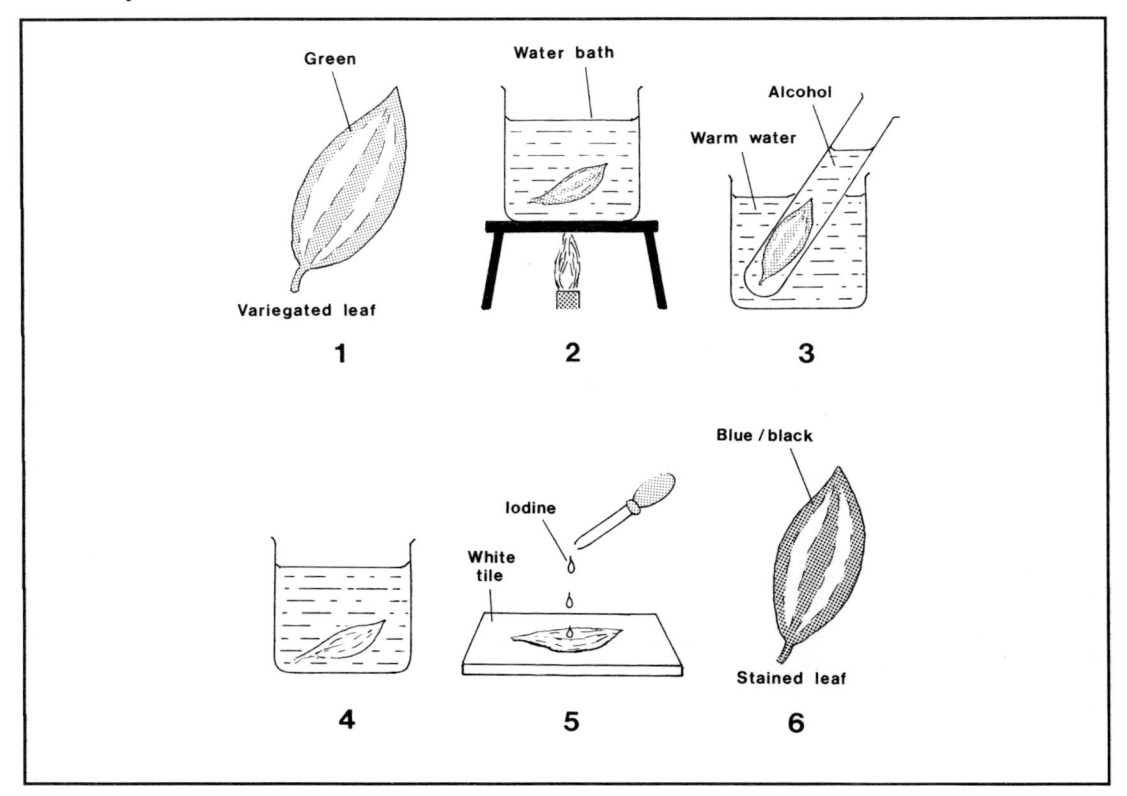

Science Work-Book 2

Results

1. Diagram of leaf at start.   Diagram of leaf at end.
   At Start: At End:

Answer the following questions:—
1. Why must you kill the leaf at the start of the experiment? ....
....
2. Alcohol is used to remove what from the leaf? ....
....
3. The Bunsen burner must be switched off before the alcohol is used. Why is this?..
....
4. Iodine solution is used to test for the presence of starch. What colour does it turn to when starch is present? ....
....
5. Explain the results of your experiment.

Science Work-Book 2

20 — Respiration

Multiple Choice Section

Questions 1-5
Each of the following questions are followed by five possible answers A,B,C,D and E. Choose the BEST answer.

1. Which of the following is the correct equation for respiration?

   A — $C_6H_{12}O_6 + O_2 \rightarrow CO_2 + H_2O$
   B — $C_6H_{12}O_6 + 6O_2 \rightarrow 6CO_2 + 6H_2O + energy$
   C — $c_6H_{12}O_6 + 6O_2 \rightarrow 6CO_2 + energy$
   D — $C_6H_{12}O_6 \rightarrow 6CO_2 + H_2O + energy$
   E — $C_6H_{12}O_6 + O_2 \rightarrow 6CO_2 + 6H_2O + energy$

2. When respiration takes place, which of the following gases is produced?
   A — Carbon dioxide
   B — Oxygen
   C — Nitrogen
   D — Carbon monoxide
   E — Hydrogen

3. Which of the following organisms respire anaerobically?
   A — Dog
   B — Plant
   C — Mouse
   D — Man
   E — Yeast

4. Anaerobic respiration results in the production of which substance?
   A — Hydrochloric acid
   B — Nitric acid
   C — Lactic acid
   D — Citric acid
   E — Sulphuric acid

5. When do the cells of a plant respire?
   A — During the day
   B — Not at any time
   C — During the night
   D — All the time
   E — In the morning

Questions 6-10 refer to the following diagram:—

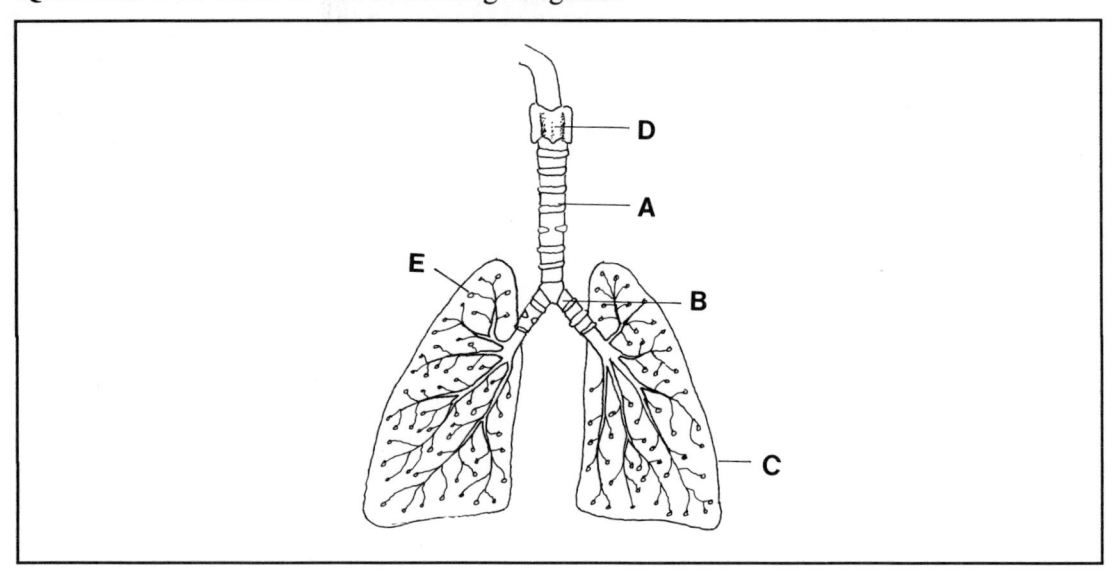

```
         D
         A
    E        B
             C
```

Choose the BEST answer from A to E

6. The correct name for this tube is the trachea. ☐

7. One of these goes to each lung. ☐

8. This is a spongy organ. ☐

9. This is where gaseous exchange takes place. ☐

10. Sounds are made in this region. ☐

Questions 11-15

In the following questions you must decide if they are TRUE or FALSE. If true put T, if false put F.

11. Breathing air in is known as expiration. ☐

12. When breathing in, the ribs are raised and the diaphragm is contracted. ☐

13. 21% of the air consists of oxygen

14. The tidal air is the maximum amount of air that can be taken in, when taking a deep breath. ☐

15. Anaerobic respiration requires oxygen. ☐

Structured Section

1. a) What is respiration? ....
....

b) What are the main differences between aerobic and anaerobic respiration? ....

c) Give an example of an aerobic organism ....

2. a) Describe how inspiration takes place.................

b) Describe how expiration takes place ....

c) What are the differences between inspired and expired air? ....

d) Describe an experiment, that shows how breathing occurs. ....

3. a) Give the equation for respiration ....

b) Where does gas exchange take place? ....

c) Explain the events that are taking place in the following diagram................

Gaseous exchange in the alveoli

Essay Section

1. Draw a labelled diagram of the human respiratory system. Explain how breathing occurs.
2. Discuss the harmful effects of smoking.
3. Explain each of the following terms: tidal air, residual air, vital capacity and total capacity. Describe an experiment to find your vital capacity.

Practical Assessment

1. **Aim:** To show that the air that we breathe out, contains more carbon dioxide than the air that we breathe in.

   Apparatus: Boiling tubes, straws, limewater, conical flasks fitted with stoppers and glass tubes arranged as follows:—

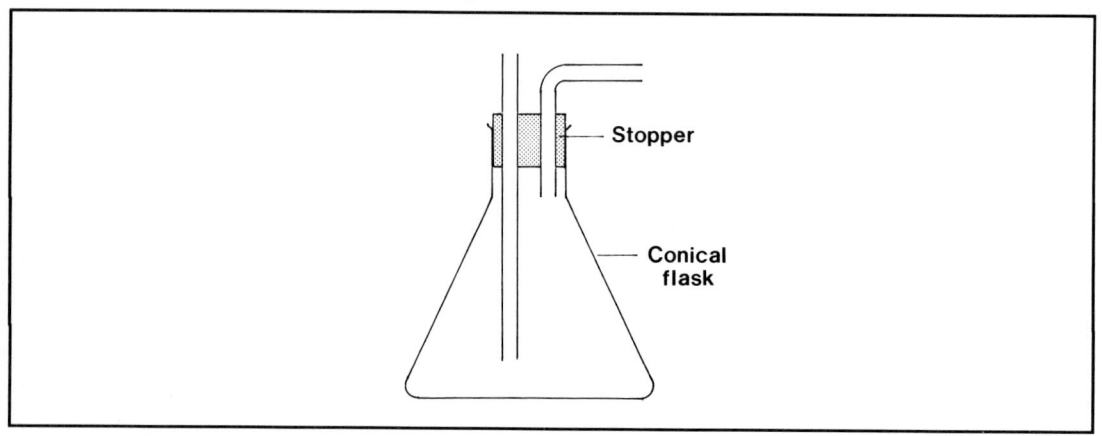

   Method

   Experiment A

   1. Pour limewater into the conical flask, so that the bottom of the longer glass tube is immersed, whilst the short tube is above the limewater.
   2. Breathe in gently, through the short tube.

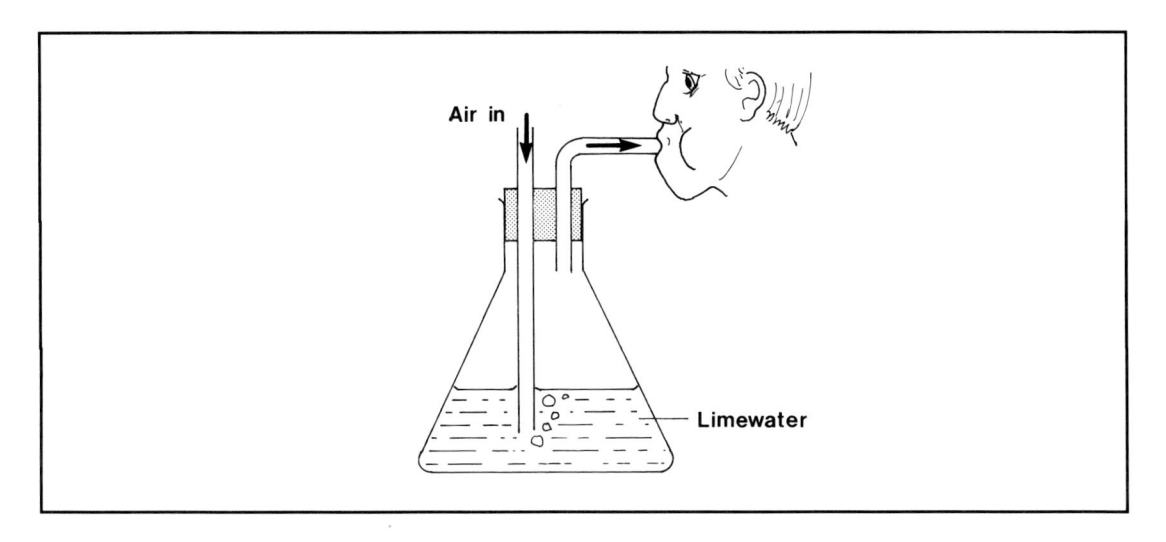

3. Note any change in the colour of the limewater.

Experiment B

1. Half fill a boiling tube with limewater.
2. Using a straw, blow gently into the limewater.

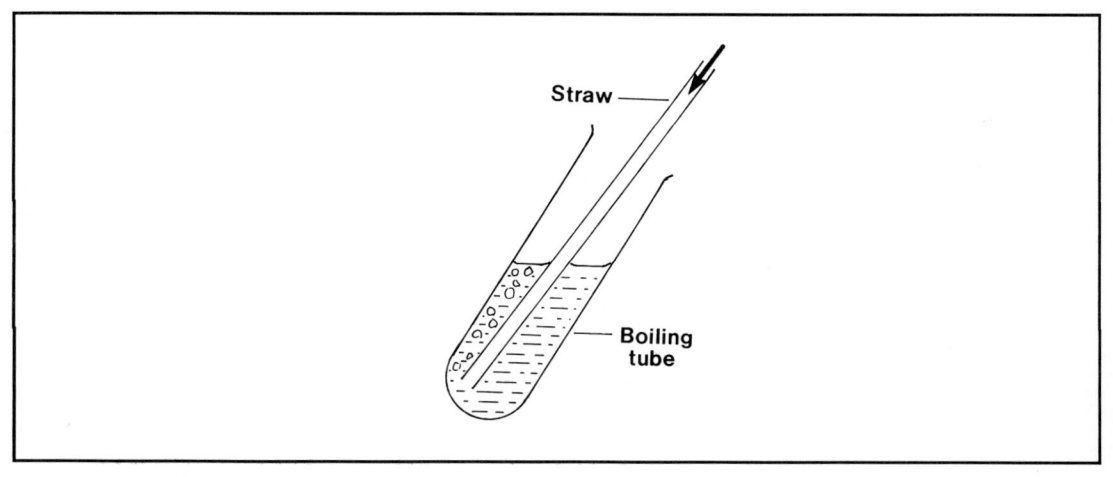

3. Note any change in the colour of the limewater.

Results

When carbon dioxide is passed through limewater, the limewater turns milky. Experiment A tests to see if there is much carbon dioxide present in the air that we breathe in. Experiment B tests exhaled air for the presence of carbon dioxide.

Experiment A		Experiment B	
Colour of limewater	CO_2 present Yes/No	Colour of limewater	CO_2 present Yes/No

Answer the following questions:

1. How do you test for carbon dioxide? ....
   ....
   ....

2. Is there much carbon dioxide in the air that we breathe in? ....
   ....
   ....

3. Does exhaled air contain carbon dioxide? ....
   ....
   ....

21 — Water and Transport in Plants

Multiple Choice Section

Questions 1-4 refer to the following diagram of an experimental apparatus.

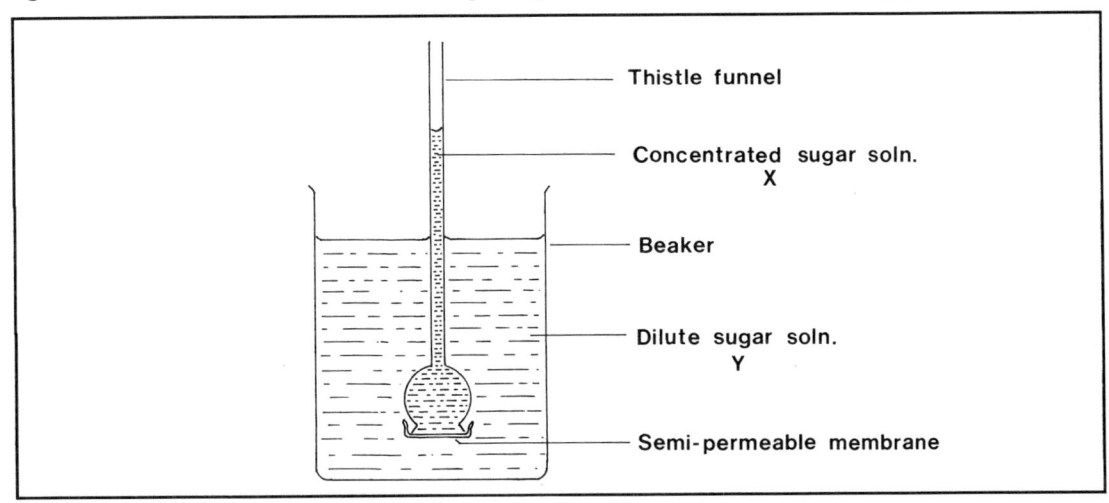

- Thistle funnel
- Concentrated sugar soln. X
- Beaker
- Dilute sugar soln. Y
- Semi-permeable membrane

Choose the BEST answer from A to E.

1. Twenty minutes after setting up the apparatus, it was found that the level of solution in the thistle funnel had risen. This was because molecules of:—
   A — Water pass from X to Y
   B — Water pass from Y to X
   C — Sucrose pass from X to Y
   D — Sucrose pass from Y to X
   E — Water and sucrose pass from Y to X ☐

2. The apparatus is used to demonstrate:—
   A — Diffusion
   B — Osmosis
   C — Respiration
   D — Absorption
   E — Photosynthesis ☐

3. The level in the thistle funnel would rise faster if the:—
   A — Temperature was decreased
   B — Atmospheric pressure was decreased
   C — Solution X was made less concentrated
   D — Solution X was made more concentrated
   E — Solution y was made more concentrated ☐

79

4. A semi-permeable membrane allows:—
   A — All molecules to pass through it
   B — Water molecules only to pass through it
   C — Large molecules to pass through it
   D — No molecules to pass through it
   E — Small molecules only to pass through it

Questions 5-10

Each question is followed by five possible answers A,B,C,D and E. Choose the BEST answer.

5. The cells of plants and animals are made up of a large proportion of water. Is it up to:—
   A — 20%   D — 80%
   B — 95%   E — 50%
   C — 90%

6. What is the name given to the tissue that transports water in plants?
   A — Xylem   D — Cortex
   B — Cambium   E — Epidermis
   C — Phloem

7. What is the name given to the loss of water vapour from the plant?
   A — Photosynthesis   D — Respiration
   B — Translocation   E — Circulation
   C — Transpiration

8. Which of the following would you use to show that a liquid is water?
   A — Litmus paper   D — Universal indicator
   B — Soda Lime   E — Cobalt chloride paper
   C — Benedict's solution

9. What is a potometer used for?
   A — Collecting small insects
   B — To measure the transpiration rate
   C — To measure the rate of photosynthesis
   D — To show translocation
   E — To demonstrate osmosis

10. Which of the following sets of conditions would result in the maximum rate of transpiration?
    A — A wet calm night   D — A dry windy day
    B — A wet windy night   E — A dry windy night
    C — A wet windy day

Questions 11-15

In the following questions you must decide if they are TRUE or FALSE. If true put T, if false put F.

11. Water is a good solvent. ☐
12. Diffusion refers to the movement of water molecules only. ☐
13. Food molecules are transported in the phloem of a plant. ☐
14. Transpiration occurs through the stomata. ☐
15. When water freezes, it becomes more dense. ☐

Structured Section

1. a) What is diffusion? ....
....
....

b) What is osmosis? ....
....
....

c) Describe the properties of water ....
....
....

d) Briefly describe an experiment that demonstrates osmosis. ....
....
....

2. The following diagram shows how water is absorbed by a root.

a) Describe how water passes from the soil surrounding the root (A) to the root hair (B) ....
....
....

b) What is the name given to the tissue that is responsible for the movement of water (D) ....
....
....

c) How does a plant obtain its supply of mineral salts? ....
....
....

d) What is root pressure? ....
....
....

Essay Section

1. What is transpiration? State the conditions which alter the rate of transpiration. Describe an experiment that demonstrates the RATE of transpiration.
2. State three reasons why a plant needs water. Describe the movement of water through a plant.

Practical Assessment

Aim: To demonstrate osmosis.

Apparatus: Beakers, visking tube, concentrated sugar solution, measuring cylinder.

Method

1. Cut a 15 cm length of visking tube and tie a knot in one end.
2. Carefully put 5 ml of concentrated sugar solution into the visking tube and tie a knot in the other end. Check that the tubing does not leak.
3. Place the filled bag into the beaker of water and leave for 30 minutes.
4. Set up a control experiment by putting 5 ml water into a similar length of visking tubing and placing in a second beaker of water and leaving for 30 minutes.
5. Remove both bags from the water, and measure the volume of liquid in each bag, by carefully cutting the bag (or untying the knot) and pouring the contents into a measuring cylinder.
6. Record your results in the table below.

Results

	Volume at Start	Volume at End
Concentrated Sugar Solution	5 ml	
Control Water	5 ml	

Questions

1. What happened to the volume of the sugar solution? ....
   ....
   ....

2. What happened to the volume of the control? ....
   ....

3. Explain your results. ....
   ....
   ....

4. What process does this experiment demonstrate? ....
   ....

22 — Blood

Multiple Choice Section

Questions 1-5
Each of the following questions is followed by five possible answers A,B,C, D and E. Choose the BEST answer from A to E.

1. How many red corpuscles are there in blood?
   A — 8,000/mm^3
   B — 10,000/mm^3
   C — 250,000/mm^3
   D — 5 million/mm^3
   E — 50,000/mm^3

2. What is the name of the red pigment found in red corpuscles?
   A — Chlorophyll
   B — Bilirubin
   C — Lymph
   D — Myoglobin
   E — Haemoglobin

3. Which of the following is NOT a characteristic of white corpuscles?
   A — Carries oxygen
   B — Has a nucleus
   C — Made in the bone marrow
   D — Produces antibodies
   E — Protects against disease

4. The average adult heart beats at rest about:—
   A — 50 times per minute
   B — 85 times per minute
   C — 72 times per minute
   D — 100 times per minute
   E — 60 times per minute

5. Which of the following are concerned with the blood clotting mechanism?
   A — Red blood cells
   B — Platelets
   C — Antibodies
   D — White blood corpuscles
   E — Lymph

Questions 6-10 refer to the following blood vessels
   A — Hepatic portal vein
   B — Aorta
   C — Pulmonary vein
   D — Renal artery
   E — Vena cava

Relate each of the descriptions below to one of the vessels above.

6. Carries blood at the greatest pressure

7. Is the only vein that carries oxygenated blood ☐

8. A vein that runs between the intestines and the liver ☐

9. Carries deoxygenated blood to the heart ☐

10. Enters the kidney ☐

Questions 11-15

In the following questions you must decide if they are TRUE or FALSE. If true put T, if false put F.

11. The liquid part of blood is called plasma. ☐

12. Arteries contain valves. ☐

13. Tissue fluid is formed when plasma leaks from blood fluid. ☐

14. Fatty acids and glycerol are carried in lymph vessels. ☐

15. Platelets produce antibodies. ☐

Structured Section

1. The following diagram shows the internal view of the heart.

a) Name the following blood vessels:
   A..
   B..
   E..
   F..
b) How many chambers are there in the heart? Give their names..............

....
....

c) Describe the pathway of blood through the heart ....
....
....

2. a) How much blood does an adult human have? ....
   b) Draw a diagram of each of the following blood cells and state their functions.
      1. Red blood cell ....
      ....
      ....

      2. White blood cell ....
      ....
      ....

      3. Platelets ....
      ....
      ....

   c) What are the four main blood groups?
      1. ....
      2. ....
      3. ....
      4. ....

3. a) Give a description of each of the following blood vessels.
      1. Arteries ....
      ....
      ....

      2. Veins ....
      ....
      ....

      3. Capillaries ....
      ....
      ....

   b) How is lymph formed? ....
   ....

   c) What are the main functions of the lymphatic system?
   ....

Essay Section

1. Give a detailed account of the composition of blood.
2. Describe the involvement of blood in a) the transport of substances, and b) defence against disease.
3. Draw a labelled diagram showing the main vessels involved in the circulation of blood. Give details of factors that might alter the rate of circulation.

Practical Assessment

Aim: To investigate the internal structure of a sheep's heart.

Apparatus: Sheep's heart, dissection board, dissection kit.

Method

1. Observe and draw the external view of the sheep's heart. Identify if possible, the following blood vessels:— pulmonary artery, pulmonary vein, aorta, vena cava.
2. Carefully cut the heart in half from top to bottom.
3. Identify the following chambers of the heart:— right and left artria, right and left ventricles.
4. Draw a diagram of the dissected heart.

23 — Excretion in Man

Multiple Choice Section

Questions 1-5 refer to the following parts of the urinary system.

    A — Renal vein
    B — Ureter
    C — Urethra
    D — Kidney
    E — Renal artery

Relate each of the descriptions below to one of the above.

1. Carries urine to the bladder ....
2. Contains blood low in urea ....
3. Urine is formed here ....
4. Carries oxygenated blood ....
5. Is the vessel that leaves the body ....

Questions 6-10

Each of the following questions are followed by five possible answers A,B,C,D and E. Choose the BEST answer from A to E.

6. Which of the following is NOT a waste product?
    A — Carbon dioxide   D — Oxygen
    B — Water   E — Salts
    C — Urea

7. The name given to the tubule where urine is formed is the:—
    A — Nephron   D — Ureter
    B — Cortex   E — Pelvis
    C — Medulla

8. Excretion is best defined as the:—
    A — Removal of unwanted food from the body.
    B — Production of urea.
    C — Elimination of urine.
    D — Removal of water from the body.
    E — Elimination of the waste products of metabolism

9. Which of the following is NOT a function of skin?
    A — Sense organ   D — Temperature regulation
    B — Respiratory organ   E — Protection
    C — Excretion

10. The part of the skin that is involved in excretion is:—
   A — Sebaceous gland   D — Hair follicle
   B — Erector muscle   E — Sense receptor
   C — Sweat gland

Questions 11-15

In the following questions you must decide if they are TRUE or FALSE. If true put T, if false put F.

11. Excretory products include urea, water and carbon dioxide. ☐

12. Plants do not excrete any waste substances. ☐

13. The kidneys produce urine. ☐

14. Urea is produced when carbohydrates are broken down. ☐

15. The skin is involved in temperature regulation. ☐

Structured Section

1. a) Define the term excretion ....
   ....
   b) What are the main waste products in man? ....
   ....
   ....
   c) Name the three main excretory organs in man.
      1. ....
      2. ....
      3. ....
   d) How does a plant excrete its waste products? ....
   ....
   ....

2. The following are all functions of the skin. Give an explanation of each one.
   a) Sense organ ....
   ....
   ....
   b) Temperature regulation ....
   ....
   ....

c) Protection ....

....

....

d) Excretion ....

....

....

Essay Section

1. Draw a labelled diagram showing the inside of a kidney. Describe the formation of urine.

2. What is meant by excretion? What part do a) lungs, b) kidneys and c) skin play in excretion?

Complete Science Work-Book 2

24 — Movement

Multiple Choice Section

Questions 1-5
Each of the following questions are followed by five possible answers A,B,C,D and E. Choose the BEST answer from A to E.

1. Movement in plants is known as:—
   - A — Locomotion
   - B — Walking
   - C — Creeping
   - D — Flowering
   - E — Tropism ☐

2. Which of the following is not a vertebral bone?
   - A — Clavicle
   - B — Sacral
   - C — Lumbar
   - D — Caudal
   - E — Cervical ☐

3. How many pairs of ribs are there in an adult man?
   - A — 14
   - B — 12
   - C — 13
   - D — 10
   - E — 15 ☐

4. Where would you find a ball and joint socket?
   - A — Elbow
   - B — Wrist
   - C — Between skull bones
   - D — Hip
   - E — Knee ☐

5. Where would you find a tendon?
   - A — Between two bones
   - B — In a joint
   - C — Between a muscle and a bone
   - D — Between two muscles
   - E — Between a bone and the skin ☐

Questions 6-10 refer to the following bones:—
   - A — Femur
   - B — Radius
   - C — Phalanges
   - D — Skull
   - E — Tibia

Relate each of the descriptions below to one of the bones above.

6. Is the longest bone in the body. ☐

7. Protects the brain ☐

8. The biceps muscle is attached to this bone ☐

9. These form the fingers of the hand ☐

10. The common name of this bone, is the shin-bone ☐

Questions 11-15

In the following questions you must decide if they are TRUE or FALSE. If true put T, if false put F.

11. The skeleton holds the body upright. ☐

12. Bones are attached to bones by tendons. ☐

13. The scapula is commonly known as the shoulder blade. ☐

14. A ball and socket joint allows movement in one direction only. ☐

15. There are more than 300 bones in the human skeleton. ☐

Structured Section

1. a) Give three functions of the human skeleton.

   1. ....
   2. ....
   3. ....

   b) List all the bones found in the arm and hand ....
   ....
   ....

   c) What is a joint? ....
   ....
   ....

   d) Name the three types of synovial joint, and give an example of each.

   1. Name ....
      Example ....
   2. Name ....
      Example ....
   3. Name ....
      Example ....

2. a) Name three types of muscle ....
   ....

b) What are muscles responsible for? ....
....
....
c) What are antagonistic muscles? Give an example ....
....
   Example: ....
d) Explain how movement of the lower arm is brought about. ....
....
....

Essay Section

1. What are the functions of the skeleton? Give the names of the bones that make up the backbone and state their particular functions.
2. Describe the three types of muscle and their functions. Give details of how muscles are involved in the movement of the body.

25 — Sensitivity

Multiple Choice Section

Questions 1-5
Each of the following questions are followed by five possible answers A,B,C, D and E. Choose the BEST answer from A to E.

1. The movement of shoots towards light is an example of:—
   A — Thigmotropism
   B — Phototropism
   C — Response
   D — Hydrotropism
   E — Geotropism

2. Which of the following is the correct pathway for a voluntary action e.g. running to catch a train?
   A — Stimulus → sensory neuron → brain→ motor neuron → response
   B — Stimulus → brain → sensory neuron → response → motor neuron
   C — Brain → stimulus → motor neuron → response
   D — Brain → sensory neuron → stimulus → motor neuron → response
   E — Stimulus → brain → sensory neuron → motor neuron → response

3. Which of the following is NOT a characteristic of the nervous system?
   A — Effect is short lived
   B — Message is carried by neurons
   C — Message is carried in the blood
   D — Message is an electrical impulse
   E — Response is usually very quick

4. The following diagram is an example of which type of cell.

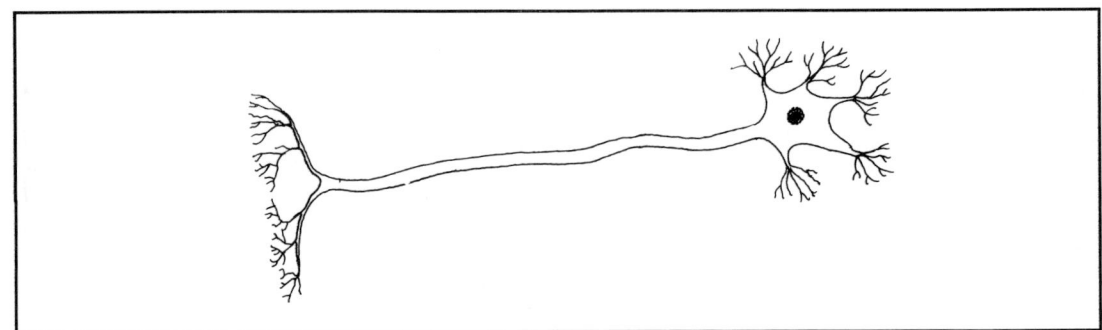

A — White corpuscle
B — Brain cell
C — Muscle cell
D — Epithelial cell
E — Neuron ☐

5. Which of these is NOT an endocrine gland?
   A — Thyroid gland
   B — Pancreas
   C — Ovaries
   D — Liver
   E — Adrenal gland ☐

Questions 6-10 refer to the following hormones:—
   A — Insulin
   B — Adrenalin
   C — Thyroxine
   D — Oestrogen
   E — Growth hormone

Relate each of the descriptions below to one of the above hormones.

6. Is produced when we are frightened. ☐

7. Produced by the pituitary. ☐

8. Controls development of secondary female characteristics. ☐

9. Diabetics suffer from a deficiency of this hormone. ☐

10. This hormone is produced in a gland found in the neck.. ☐

Questions 11-15

In the following questions you must decide if they are TRUE or FALSE. If true put T, if false put F.

11. Plants do not respond to stimuli. ☐

12. Nerves are made of bundles of neurons. ☐

13. Sensory neurons carry messages from the receptor to the brain. ☐

14. The pituitary gland is sited just below the brain. ☐

15. A hormone can be defined as a chemical messenger. ☐

Structured Section

1. a) What is a tropism? ....
   ....
   ....

   b) Give three examples of positive tropisms together with their correct names.

     1. Positive tropism ....

Results

	Bean A	Bean B	Bean C
At Start			
After one week			

Questions

1. Why did you fill the beaker with damp peat? ....
   ....
   ....

2. Why was it important not to let the peat dry out? ....
   ....
   ....

3. What happened to the size of the beans? ....
   ....
   ....

4. Describe the growth of the roots and shoots in each beaker
   Beaker A ....
   ....
   Beaker B ....
   ....
   Beaker C ....
   ....

26 — The Sense Organs

Multiple Choice Section

Questions 1-5 refer to the following diagram of a section through the eye of a mammal. Choose the correct answer from A to E.

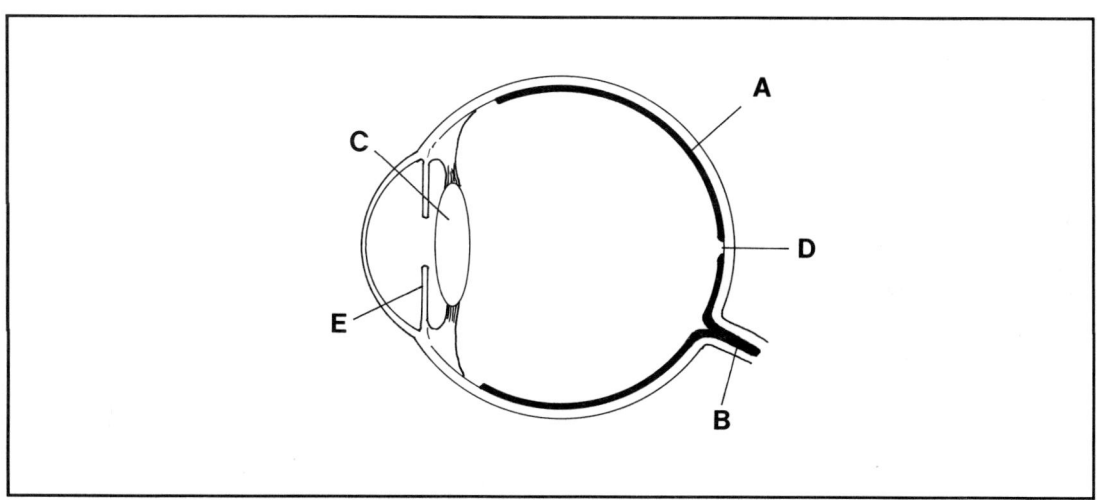

1. This part of the eye gives it, its colour. ☐
2. Is responsible for focusing light. ☐
3. Contains light sensitive cells. ☐
4. Carries messages to the brain. ☐
5. Place of most accurate vision. ☐

Questions 5-10 refer to the following parts of the ear.
    A — Pinna
    B — Ear bones
    C — Ear drum
    D — Auditory nerve
    E — Semi-circular canals

Relate each of the following statements below to one of the above parts.

6. Carry vibrations from the eardrum to the oval window. ☐
7. Are concerned with balance. ☐
8. Vibrates when sound waves hit it. ☐

9. Transmits nerve impulses from the ear to the brain. ☐

10. Gathers sound waves. ☐

Questions 11-15

In the following questions you must decide if they are TRUE or FALSE. If true put T, if false put F.

11. Taste buds are only sensitive to sweet, sour and salt tastes. ☐

12. Our sense of taste is linked to our sense of smell. ☐

13. Skin is a sense organ. ☐

14. Ears are responsible for hearing only. ☐

15. The optic nerve carries nerve impulses from the brain. ☐

Structured Section

1. a) What are our five senses?

   1 ....
   2 ....
   3 ....
   4 ....
   5 ....

   b) What are the four types of sensory receptor found in human skin?

   1 ....
   2 ....
   3 ....
   4 ....

   c) Suggest reasons why most of our sense organs are situated on our head? ......
   ....
   ....

2. a) What are the functions of the following parts of the eye?

   1. Lens ....
   ....

   2. Retina ....
   ....

   3. Conjuctiva ....
   ....

4. Optic nerve ....
....

b) The following diagrams show the external view of the eye at different times of the day.

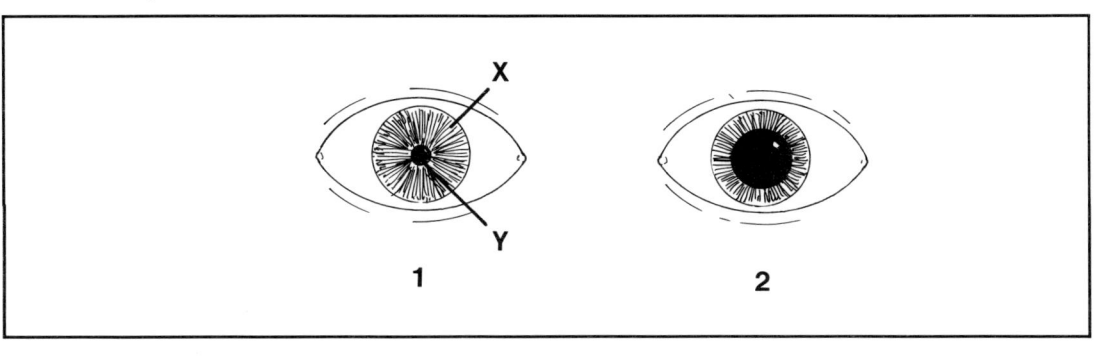

(i) Which diagram represents the eye at night? ....
....
(ii) Explain your answer ....
....
(iii) Give the names of
  X ....
  Y ....

Essay Section

1. Draw a labelled diagram of the ear. Describe how sound waves pass through the ear to the cochlea.
2. What are our main sense organs. Briefly describe the information that they give us about our environment.

Practical Assessment

Aim: To discover if two eyes are better than one.

Apparatus: Mm graph paper, coloured pencils.

Method

1. Mark 20 small crosses at random on the graph paper.
2. Cover your left eye and using a coloured pencil, **quickly** stab each of the 20 crosses.
3. Cover your right eye and, using a different coloured pencil, **quickly** stab each of the 20 crosses again.
4. Repeat this with both eyes open and using a third colour.
5. Mark your 'stabs' according to the following table.

	Marks
Within 1 mm	5
Between 1-2 mm	4
Between 2-3 mm	3
Between 3-4 mm	2
Between 4-5 mm	1
> 5 mm	0

6. Complete the results table.

Cross	Marks			Cross	Marks		
	Left Eye Open	Right Eye Open	Both Eye Open		Left Eye Open	Right Eye Open	Both Eye Open
1				11			
2				12			
3				13			
4				14			
5				15			
6				16			
7				17			
8				18			
9				19			
10				20			
Total				Total			

7. Work out the total marks for each set of results.

Questions

1. Did you score higher marks with your left eye open, your right eye open, or both eyes open? ....
   ....

2. From your results, do you think that two eyes are better than one at judging distances ....
   ....

3. Why did you use 20 crosses and not just 1? ....
   ....

4. Can you think of any other reason why two eyes are better than one? ....
   ....

Complete Science Work-Book 2

27 — Cycles in Nature

Multiple Choice Section

Questions 1-5

Each question is followed by five possible answers A,B,C,D and E. Choose the BEST answer from A to E.

1. Nitrogen makes up which percentage of the air?
   - A — 20%
   - B — 22%
   - C — 70%
   - D — 75%
   - E — 78%

2. What is the name given to the bacteria that live in root nodules in plants?
   - A — Putrifying bacteria
   - B — Denitrifying bacteria
   - C — Nitrifying bacteria
   - D — Nitrogen-fixing bacteria
   - E — Nitrogen-losing bacteria

3. Which of the following, results in the loss of carbon dioxide in the air?
   - A — Photosynthesis
   - B — Combustion
   - C — Transpiration
   - D — Respiration
   - E — Translocation

4. Which of the following is NOT a fossil fuel?
   - A — Gas
   - B — Electricity
   - C — Oil
   - D — Coal
   - E — Peat

5. Which of the following substances contains nitrogen?
   - A — Carbohydrate
   - B — Vitamins
   - C — Protein
   - D — Fat
   - E — Mineral Salts

Questions 6-15

In the following questions you must decide if they are TRUE or FALSE. If true put T, if false put F.

6. Plants and animals can absorb nitrogen from the air.

7. Herbivorous animals obtain their protein by eating plants.

8. Nitrifying bacteria convert ammonia to nitrite and nitrate.

103

9. Lightning creates a high enough temperature to bring about a chemical reaction between nitrogen and oxygen, forming nitrogen oxides. ☐

10. Carbon dioxide gas makes up about 20% of the air. ☐

11. Respiration has the equation:—
$C_6H_{12}O_6 + 6O_2 \rightarrow 6CO_2 + 6H_2O + energy.$ ☐

12. Transpiration is of biological importance in the water cycle. ☐

13. Fossil fuels are composed of the remains of dead plants and animals. ☐

14. Plants make their own food by translocation. ☐

15. Purifying bacteria convert nitrates into free nitrogen. ☐

Structured Section

1. a) Name three substances that are involved in cycles in nature.

   1. ....
   2. ....
   3. ....

   b) How is nitrogen in the air converted into nitrate for plants ....
   ....
   ....

   c) What do denitrifying bacteria do? ....
   ....
   ....

   d) Why is lightning important in the nitrogen cycle? ....
   ....
   ....

2. a) How, and in which cycles are the following processes important?
   A Photosynthesis ....
   ....

   Cycle ....
   ....

   B Respiration ....
   ....

   Cycle ....
   C Transpiration ....
   ....

Cycle ....
D Combustion ....
....
Cycle ....

Essay Section

1. Draw a diagram to represent the nitrogen cycle. Discuss the involvement of bacteria in this cycle.
2. The carbon cycle involves the gain and loss of carbon dioxide in the air. Discuss the biological processes involved.

28 — Ecology

Multiple Choice Section

Questions 1-5 refer to the following groups:—
- A — Herbivores
- B — Decomposers
- C — Omnivores
- D — Producers
- E — Carnivores

Refer each of the descriptions below to one of the above groups.

1. These organisms feed on dead plants and animals.

2. Plants belong to this group.

3. Man is a member of this group.

4. These are animals that eat meat only.

5. These animals feed on the producers only.

Questions 6-10

Each of the following questions are followed by five possible answers, A,B,C,D and E. Choose the BEST answer from A to E.

6. Which of the following environmental factors is not climatic?
   - A — Soil pH
   - B — Rainfall
   - C — Light intensity
   - D — Wind
   - E — Temperature

7. Acid rain is caused by which pollutant?
   - A — Fertilizers
   - B — Sewage
   - C — Oil
   - D — Sulphur dioxide
   - E — Smoke

8. A pooter is used to collect:—
   - A — Butterflies
   - B — Insects
   - C — Small mammals
   - D — Plankton
   - E — Limpets

9. Which of the following shows a correct food chain
   - A — Greenfly → roseleaves → ladybird
   - B — Ladybird → greenfly → roseleaves
   - C — Roseleaves → greenfly → ladybird
   - D — Ladybird → roseleaves → greenfly
   - E — Roseleaves → ladybird → greenfly

10. Which group of organisms are present in the largest numbers?
    A — Herbivores
    B — Decomposers
    C — Scavengers
    D — Carnivores
    E — Producers

Questions 11-15

In the following questions you must decide if they are TRUE or FALSE. If true put T, if false put F.

11. The place an animal lives in, is known as its habitat.

12. Edaphic factors are factors related to the climate.

13. A Tullgren funnel is used to extract small animals from soil.

14. Carnivores are primary consumers.

15. Most fungi are decomposers.

Structured Section

1. a) Explain each of the following terms:—
   1. Ecosystem ....
   ....
   ....
   2. Habitat ....
   ....
   ....
   3. Food chain ....
   ....
   ....
   4. Competition ....
   ....
   ....
   5 Colonisation ....
   ....
   ....
   b) What are producers? ....
   ....
   ....

c) List the FOUR types of consumers, and give an example for each one.

1. ....
Example ....
2. ....
Example ....
3. ....
Example ....
4. ....
Example ....

2. The following diagram represents a food web.

```
        MAN  ←────────→  SEALS
          ↖   ↗  ↖       ↑
           ╲ ╱    ╲      │
        MACKEREL   HERRING
            ↖       ↗  ↖
             ╲     ╱    │
          SMALL CRUSTACEANS
                  ←
                      ╲
                       ALGAE
```

a) Name the producers in this food web ....
....
b) What does a herring feed on? ....
....
c) Give the names of any secondary consumers ....
....
d) Describe man's position in this food web ....
....
....

3. a) What is pollution? ....
....
....

b) Give examples of the three main water pollutants and state their effect on the environment.

1. ....

....
....
2. ....
....
....
3. ....
....
....

c) What is meant by conservation? ....
....

Essay Section

1. Describe how you would collect a random sample of animals from a named habitat. Give details of any nets and/or traps that you might use.
2. Describe the plants and animals found in a woodland habitat, that you have studied.
3. What is ecology? Give details of the interdependence of plants and animals.

Practical Assessment

Aim: To study a leaf litter environment.

Apparatus: Box for leaf litter collection, magnifying lens, petri dishes, simple leaf keys and animal keys or other identification aids.

Method

1. Collect a sample of leaf litter, include some surface soil.
2. Sort out all animals from the leaf litter, and use your knowledge of classification to identify them. (Use keys etc. if available).
3. Identify the leaves in your leaf litter sample.
4. Draw diagrams of all plants and animals found, and give their names.
5. Return animals to their correct habitat.

29 — Genetics

Multiple Choice Section

Questions 1-5

Each of the following questions have five possible answers A,B,C,D and E. Choose the BEST answer from A to E.

1. Which of the following characteristics is NOT inherited?
   - A — Eye colour
   - B — Curly hair
   - C — Ability to ride a bike
   - D — Freckles
   - E — Ability to roll your tongue

2. How many chromosomes are there in an adult man?
   - A — 54
   - B — 30
   - C — 25
   - D — 46
   - E — 23

3. How many chromosomes are there in a female ovum?
   - A — 54
   - B — 30
   - C — 25
   - D — 46
   - E — 23

4. Which of the following is an inherited disease?
   - A — Influenza
   - B — Haemophilia
   - C — German measles
   - D — Polio
   - E — Whooping cough

5. The genetic make-up of a cell is known as:—
   - A — Genotype
   - B — Gene
   - C — Chromosome
   - D — Homozygous
   - E — Phenotype

Questions 6-10 refer to the following terms:—
   - A — Dominant
   - B — Phenotype
   - C — Genotype
   - D — Homozygous
   - E — Heterozygous

Relate each of the following statements to one of the above terms.

6. The physical expression of a gene. ☐

7. Both genes for a characteristic are identical. ☐

8. When two different genes are present. ☐

9. When one gene masks the other gene, it is said to be:— ☐

10. This is the type of genes an organism has. ☐

Questions 11-15

In the following questions you must decide if they are TRUE or FALSE. If true put T, if false put F.

11. Freckles are an inherited characteristic. ☐

12. Blue eyes are dominant over brown eyes. ☐

13. Two dark haired parents cannot have a fair-haired child. ☐

14. Genetic information is carried on chromosomes. ☐

15. Chromosomes are present in the nucleus of every living human cell. ☐

Structured Section

1. a) What is genetics? ....
   ....

   b) Give THREE characteristics that are inherited?
      1. ....
      2. ....
      3. ....

   c) What is meant by a dominant characteristic? ....
      ....
      ....

   d) How many chromosomes are present in man? ....
      ....

2. Brown eyes are dominant over blue eyes. The diagram below shows what happens when a homozygous brown-eyed man marries homozygous a blue-eyed woman, and they have children.

```
PARENTS:      (BB)                    (bb)

GAMETES:    B      B            b        b

F₁
CHILDREN:  (Bb)   ( )          ( )     (Bb)
            X              Y
```

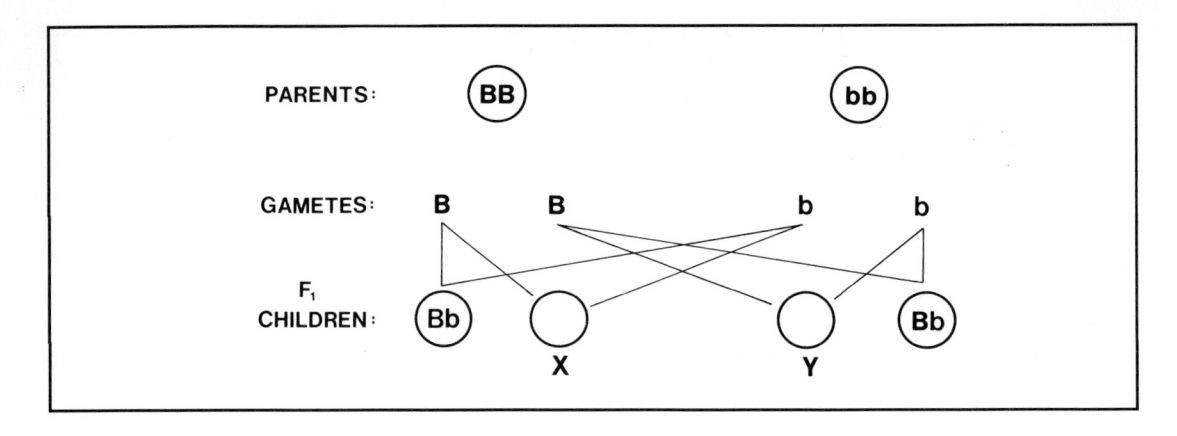

a) Complete the missing genotypes X and Y ....
....

b) What colour eyes will the children have? ....
....

c) Will they be homozygous or heterozygous for eye colour? ....
....

Essay Section

1. Give details of how genetic information is carried from parent to child. Include the following terms in your answer:— gene, chromosomes, dominant, recessive, phenotype and genotype.

Complete Science Work-Book 2

30 — Force

Multiple Choice Section

Questions 1-5 concern the following
- A — Weight
- B — Forces
- C — Newton
- D — Mass
- E — Stretching

1. This is the force of gravity on a body

2. This does not change when a body moves from one place to another.

3. This is the unit of force

4. These always occur in pairs

5. This property of a body is dependent on the distance from the centre of the earth

Questions 6-10

For each of the questions ONE or MORE of the answers are correct. Decide which of the responses is (are) correct. Then choose from

A	B	C	D	E
1,2,3 correct	1,2 only	2,3 only	1 only	3 only

6. Which of the following can change when a force is exerted?
   - 1 — Speed
   - 2 — direction
   - 3 — shape

7. To every action there is an equal and opposite
   - 1 — Mass
   - 2 — Force
   - 3 — Reaction

8. The weight of a body depends on
   - 1 — Mass
   - 2 — Gravity
   - 3 — Speed

113

9. Three similar elastic bands are tied as shown

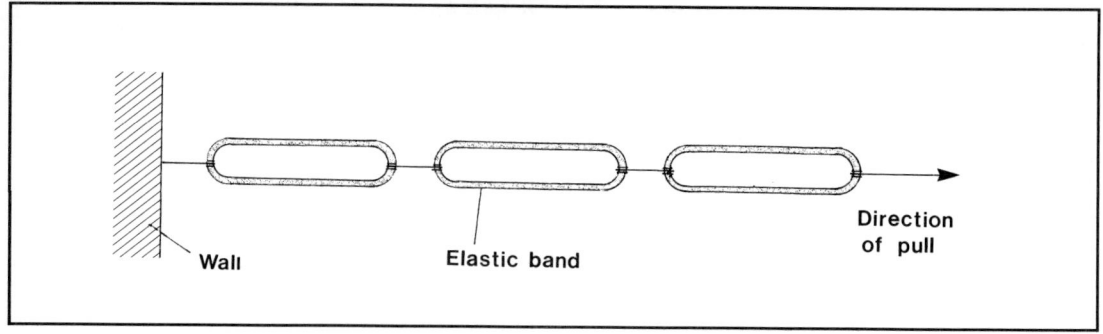

Which of the following is true?
   1. They all stretch by the same amount
   2. The total extension is three times the extension of one band
   3. The band in the centre is the only one which stretches ☐

10. The weight and mass of an astronaut were both measured on the Moon and on the Earth. Which of the following is true?
   1. The weight was more on the Moon than on the Earth
   2. The weight was less on the Moon than on the Earth
   3. The mass was the same both on the Moon and on the Earth. ☐

Questions 11-15

In the following questions you must decide if they are TRUE or FALSE. If true put T, if false put F.

11. Power is defined as the rate at which work is done. ☐

12. The unit of both work and energy is the Joule. ☐

13. The amount of work done is calculated by multiplying force and distance moved together. ☐

14. The relationship between the extension of a spring and the load placed on it is called Newton's Law. ☐

15. A Newton is approximately the weight of an apple. ☐

Structured Section

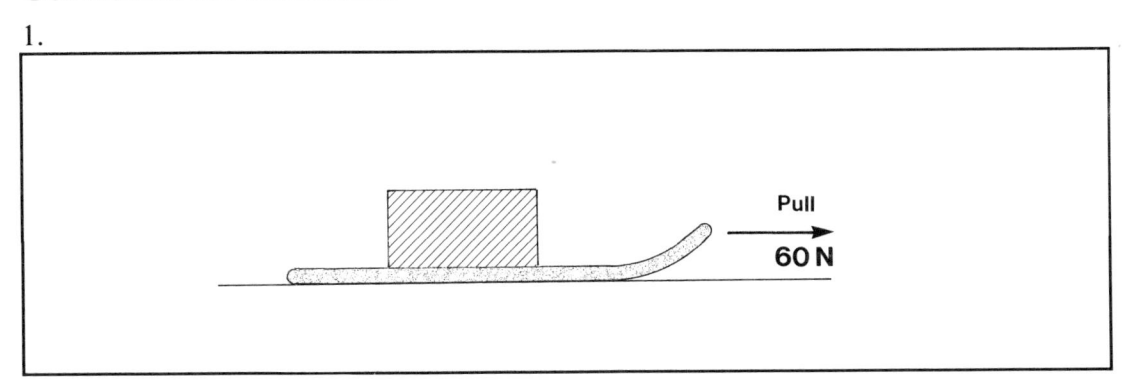

1.

A boy pulls a sled, carrying a box of mass 20 kg, with a constant force of 60N.

a) What is the weight of the box ....

b) Weight is a force. On the diagram, draw an arrow to show the direction of this force. Label it W

c) There is a frictional force between the sled and the ground. On the diagram, draw an arrow, labelled F, to show its direction

d) If the frictional force has a value of 20N, what is the resultant (resulting) force on the sled ....

e) If the boy pulls the sled a distance of 20m, how much work does he do? ....
....

2. In an experiment to find how a steel spring stretches with the load applied, the following readings were obtained

Load (N)	Extension (mm)
1.0	1.3
1.5	2.0
2.0	2.7
3.0	4.0
3.5	4.6
4.0	5.3
5.0	6.7

On the graph paper below plot the values given and using the points, draw a line to show how the spring behaves when stretched.

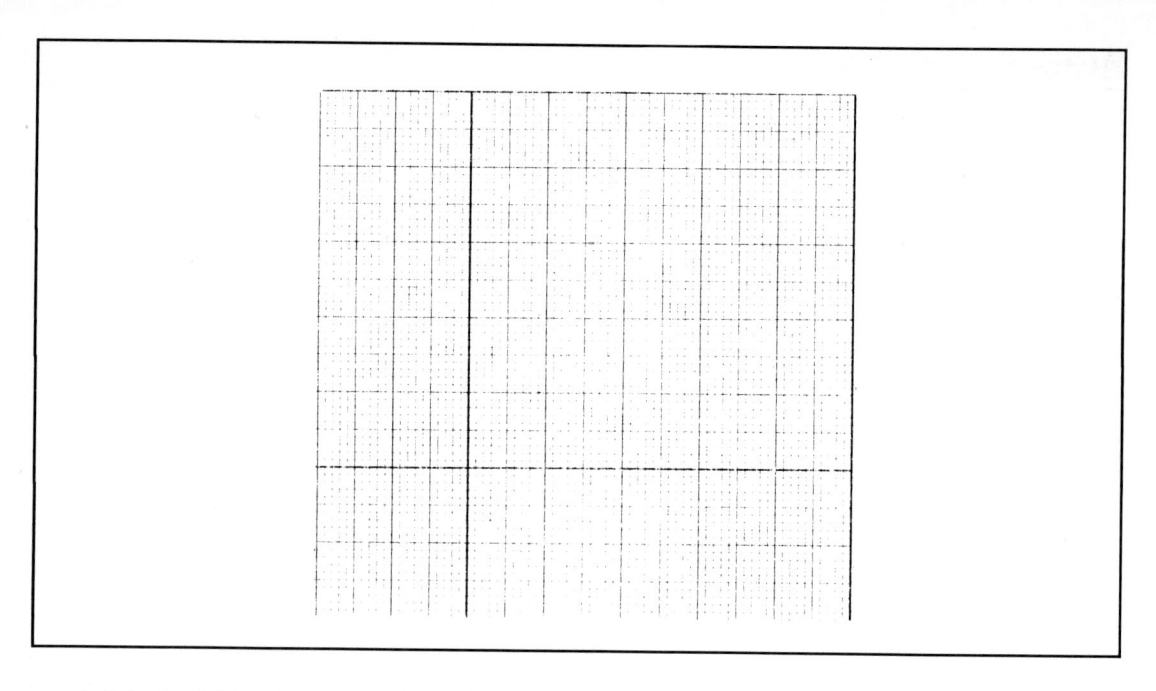

a) What will be the extension when a load of 2.5N is placed on the spring? ......

b) When an unmarked mass is placed on the spring, it extends 8mm.
   (i) What force does it exert on the spring? ....
   (ii) What is the value of the mass in grams? ....

Essay Section

1. Describe an experiment you have performed which shows that forces always occur in pairs.
2. Write an essay about the forces and their effects in everyday life e.g. the force you exert on a chair when you sit on it and the force it exerts on you.

Practical Assessment

Aim: To investigate the force registered by two springs when they are in parallel or series supporting a known weight.

Arrangement of springs

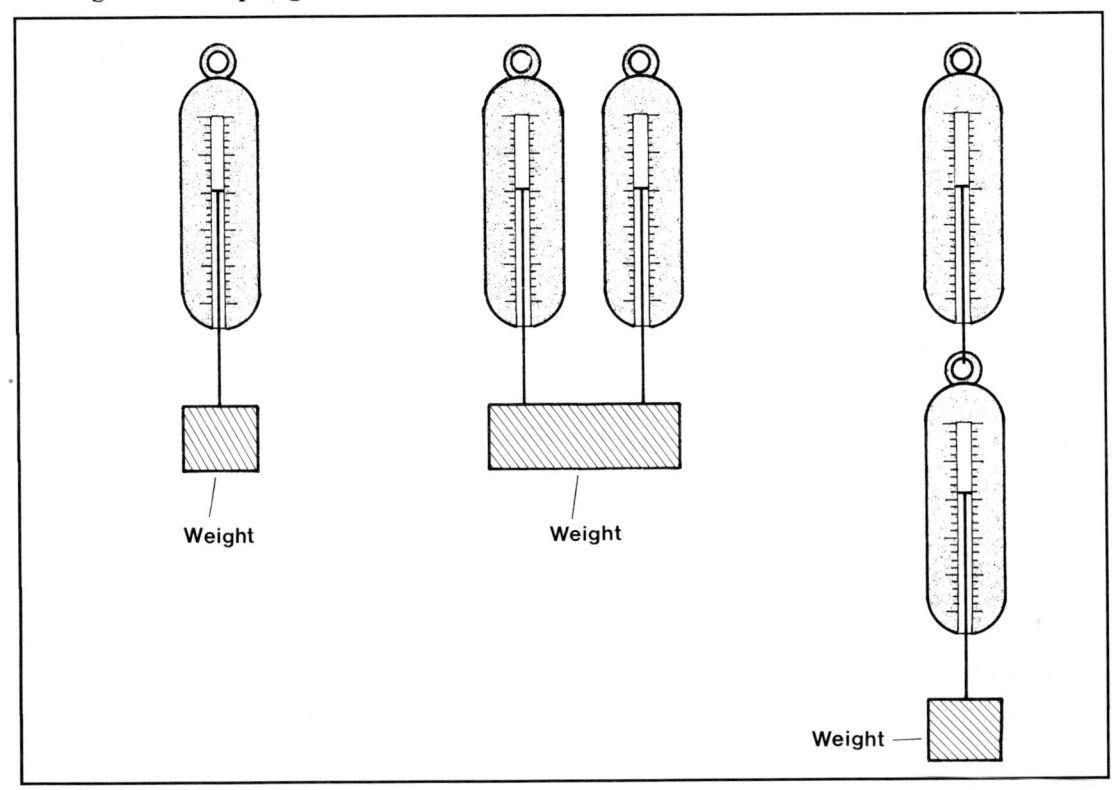

Extra Apparatus

4 blocks of different weights with two supporting loops attached to them and labelled 1,2,3,4.

Method

1. Copy the results table below onto a piece of paper. Write your conclusion underneath the results table.
2. Place each block on a single spring balance in turn and write its weight in the table.
3. Support the two balances from the clamp stands and hang each weight onto them in turn. Enter your results in the table.
4. Finally arrange two balances as shown in 3, and suspend each weight in turn from them. Take the readings of each balance and enter in the results table.

Block No.	Reading on single balance	Reading on each spring balance in parallel	Reading on each spring balance in series

5. What conclusions can you draw from your results?

Complete Science Work-Book 2

31 — Turning Forces

Multiple Choice Section

Questions 1-5 concern the following

    A — Moment
    B — Centre of Gravity
    C — Equilibrium
    D — Levers
    E — Fulcrum

1. A body which is balanced and not moving is in ☐

2. This is the turning effect of a force ☐

3. For the principle of moments to apply to a body, it must be in this state ☐

4. The point in a body through which its weight acts no matter what its position ☐

5. A crowbar, door handle and your arm are examples of these ☐

Questions 6-10

Each question is followed by five possible answers A,B,C,D and E. Choose the BEST answer.

6. Racing cars are designed to have:—
    A — low centre of gravity   D — neutral equilibrium
    B — high centre of gravity   E — balanced equilibrium
    C — zero moment   ☐

7. The moment of force is given by:—
    A — Force/Speed   D — Force × perpendicular
    B — Force × Speed   distance
    C — Force/perpendicular   E — Force × speed ×
         distance   perpendicular distace ☐

8.

[Diagram showing a hand holding a ruler vertically, labeled "Ruler"]

The ruler in the diagram is said to be in:—
  A — Stable equilibrium   D — Stationary equilibrium
  B — Unstable equilibrium   E — Balanced equilibrium
  C — Neutral equilibrium   ☐

9. A class 2 lever has:—
  A — the load between the effort and the fulcrum
  B — the load and the fulcrum at the same position
  C — the load and the effort on either side of the fulcrum
  D — the effort between the load and the fulcrum
  E — the effort and the fulcrum at the same position   ☐

10. Four identical discs are placed on a ruler as shown in the diagram. Where must a fifth disc be placed on the right hand side in order to balance the ruler?

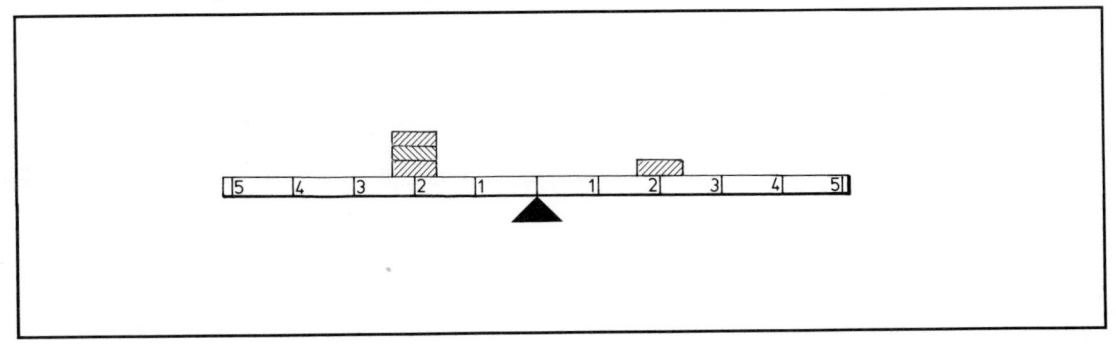

[Diagram of a ruler balanced on a fulcrum, marked 5 4 3 2 1 1 2 3 4 5, with discs stacked at position 2 on the left and a disc at position 2 on the right]

  A — 1   D — 4
  B — 2   E — 5
  C — 3   ☐

119

Questions 11-15

In the following questions you must decide if they are TRUE or FALSE. If true put T, if false put F.

11. The forearm is an example of a class 1 lever. ☐

12. A steelyard is a type of electronic balance. ☐

13. The principle of moments states that the sum of the clockwise moments is equal to the sum of the anticlockwise moments when a body is in equilibrium. ☐

14. The centre of gravity of a uniform metre rule is at the 50 cm mark. ☐

15. If a body is in stable equilibrium, disturbing it raises its centre of gravity. ☐

Structured Section

1.

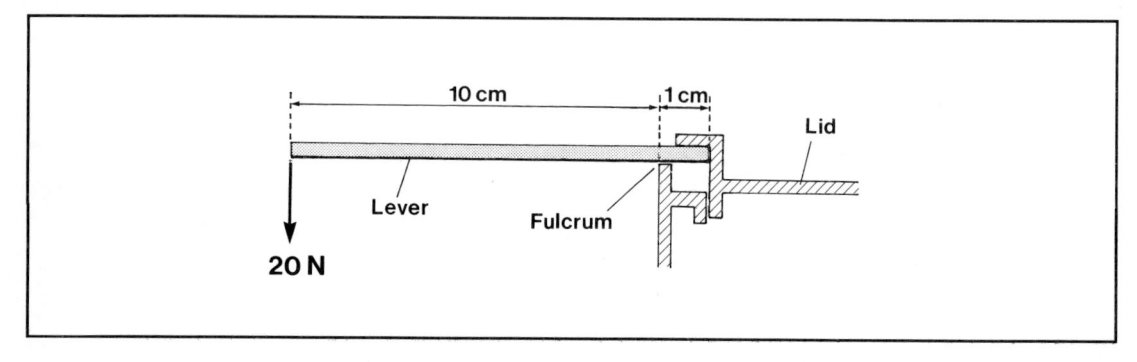

The diagram shows a paint tin being opened using a lever.

a) What class of lever is this? ....

b) Draw on the diagram an arrow to show the direction of the force exerted on the lid by the lever. Label it F. ....

c) Using the principle of moments calculate the force F exerted on the lid by the lever ....

d) State two changes you could make to increase the size of F, if it is not large enough to lift the lid. ....

2. Bearing in mind that when a body is in equilibrium, the total upward force on it is equal to the total downward force, study the diagram below and then answer the questions.

The metre rule has a weight of 1N.

```
            10       40                    90
      ─────────┬──────┼──────────────────────┬─────
               │      │Fulcrum               │
              ┌┴┐                           ┌┴┐
              │ │                           │ │
              └─┘                           └─┘
              5 N                           X N
```

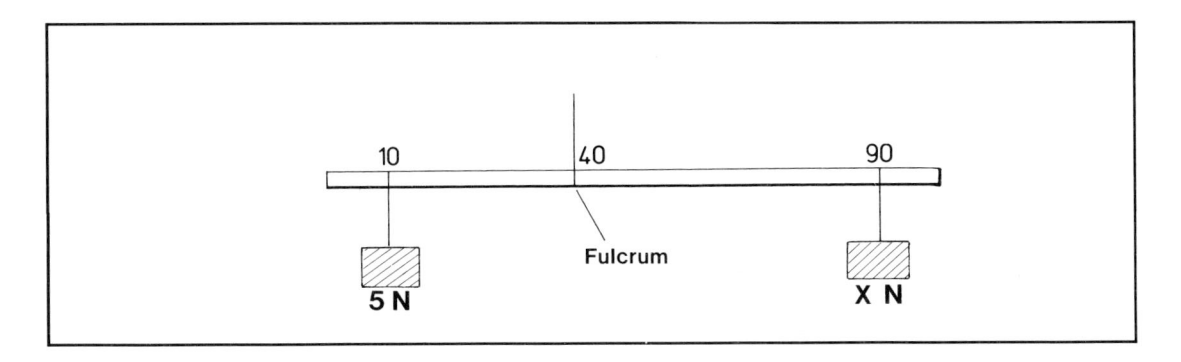

a) Draw an arrow to show the position and direction of the weight of the rule. Label it W.

b) Using the principle of moments calculate the value of X.....................

c) What is the total downward force? ....

d) What is the tension (force) in the string?...................................

Essay Section

1. Describe an experiment you have performed to find the weight of a metre rule. Include in your account a diagram, method and a description of how the weight was calculated.

2. Discuss the different classes of lever and draw three examples of each type, clearly labelling the effort, load and fulcrum (pivot).

Practical Assessment

Aim: To investigate how the applied force varies with the distance to balance a constant moment.

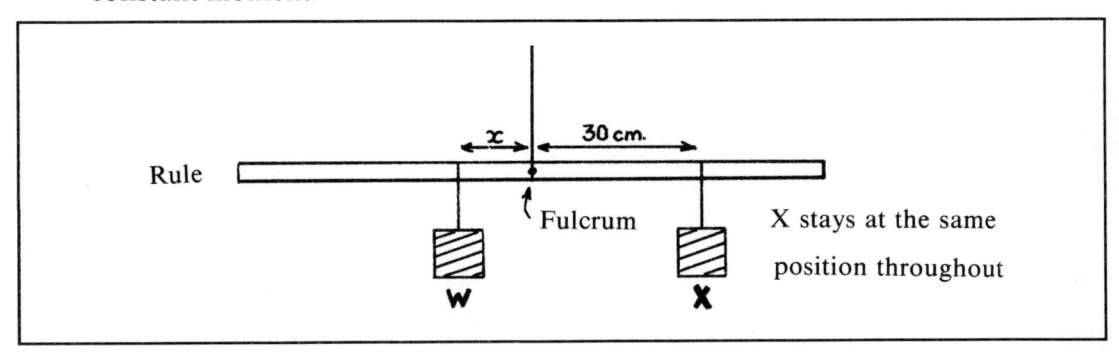

```
                    ← x →  ← 30 cm. →
   Rule  ┌──────────────┬──────────────┐
         └──────────┬───┴───┬──────────┘
                    │Fulcrum│
                   ┌┴┐     ┌┴┐        X stays at the same
                   │ │     │ │        position throughout
                   └─┘     └─┘
                    W       X
```

Method

1. Set up the apparatus as shown above placing a 6N weight at a distance 10 cm from the fulcrum so the ruler balances.. This means that x = 10 cm.

Complete Science Work-Book 2

2. Note the value of the weight and the distance x in a table similar to that below.
3. Now using a 5N weight repeat the experiment and note the results.
4. Continue to balance the ruler with 4N, 3N and 2N weights in turn.
5. Plot a graph of distance against force. Can you decide how the distance from the fulcrum varies with the force?

Force /N	Distance x/cm
6	
5	
4	
3	
2	

Complete Science Work-Book 2

32 — Motion

Multiple Choice Section

Questions 1-5 concern the following
- A — Distance
- B — Velocity
- C — Acceleration
- D — Displacement
- E — Speed

1. Distance travelled in unit time ☐
2. Distance travelled in a particular direction ☐
3. The change of velocity in unit time ☐
4. When divided by time this gives the average speed ☐
5. Distance travelled in a particular direction in unit time ☐

Questions 6-10

For each of the questions ONE or MORE of the answers are correct. Decide which of the responses is (are) correct. Then choose

A	B	C	D	E
1,2,3 correct	1,2 only	2,3 only	1 only	3 only

6. Which of the following are scalar quantities?
    1. Displacement
    2. Area
    3. Speed ☐

7. Which of the following are vector quantities?
    1. Displacement
    2. Velocity
    3. Speed ☐

8. Acceleration occurs when
    1. Velocity decreases
    2. Velocity increases
    3. Direction changes ☐

9. The velocity of a car changes by 10 m/s in 25 secs. What is its acceleration?
    1. 250 m/s^2
    2. 4.0 m/s^2
    3. 0.4 m/s^2 ☐

10. A person walks 6 km in 50 minutes. What is their average speed in m/s?
    1. 120.0 m/s
    2. 12.0 m/s
    3. 2.0 m/s

Questions 11-15

In the following questions you must decide if they are TRUE or FALSE. If true put T, if false put F.

11. A scalar quantity is one which has no direction.

12. Length is a vector quantity.

13. The unit of velocity is m/s.

14. m/s² is read as metres per two seconds.

15. A car going round a corner at 30 mph can be said to be accelerating.

Structured Section

1. An aircraft leaves London Heathrow Airport at 10.00 a.m. and flies North 10° West to Newcastle Airport arriving at 10.50 a.m. The distance from London to Newcastle is 400 km.
    (i) Find its average speed in m/s for the complete journey ....
    (ii) Does the aircraft, when on course for Newcastle, move with a velocity or speed? ....
    (iii) If the aircraft reaches a velocity of 250 km/hr on take off after 30 seconds, what is its acceleration? ....

2. Rearrange the following in order, with the slowest at the top and the fastest at the bottom

list	arranged list
Rocket (Saturn) Cheetah Bicycle Ostrich Snake Train (Inter-City) Airliner	

Essay Section
1. Using reference books, write an account of how transport has changed since 1850 to the present day, indicating how journey time has shortened.
2. Make a list of all the situations where acceleration occurs.

Practical Assessment
Aim: To measure the velocity of a bicycle.

Instructions

Devise an experiment to find the velocity of a bicycle. Write an account which lists the apparatus you would use, how you would carry out the experiment, the readings you would take and from these how you would find the velocity.

Complete Science Work-Book 2

33 — Pressure

Multiple Choice Section

Questions 1-5

For each of the questions ONE or MORE of the answers are correct. Decide which of the responses is (are) correct. Then choose

A	B	C	D	E
1,2,3 correct	1,2 only	2,3 only	1 only	3 only

1. Atmospheric pressure can be measured using
   1. Mercury barometer
   2. Aneroid barometer
   3. Fortin barometer ☐

2. The pressure exerted by a liquid on the bottom of a beaker depends on
   1. Depth of liquid
   2. Density
   3. Area of base ☐

3. The corrugated metal box of an aneroid barometer contains
   1. Air at atmospheric pressure
   2. Vacuum
   3. Partial vacuum ☐

4. When a stone is placed in water it experiences an upward force, this is known as the
   1. Upthrust
   2. Buoyancy
   3. Fluid ☐

5. Ships float because:—
   1. They are lighter than water
   2. The average density is less than that of water
   3. The weight of water displaced is equal to their own weight ☐

Questions 6-10

Each question is followed by five possible answers A,B,C,D and E. Choose the BEST answer.

6. Pressure is defined as:—
   A — Force × depth   D — Force/Area
   B — Force × area   E — Area/Depth
   C — Area/Force   ☐

126

7. What is the pressure exerted by the block on the bench? Weight of block 10 N.

[Diagram of block: 2 cm × 5 cm × 10 cm]

    A — 1N/cm²   D — 0.2N/cm²
    B — 0.5N/cm²   E — 20N/cm²
    C — 100N/cm²

8. When a can containing water is heated to boiling, then sealed and allowed to cool. The can
    A — Does not change shape   D — Expands
    B — Blows up   E — Contracts
    C — Collapses

9. Atmospheric pressure can be written as
    A — 100 cm of Hg   D — 76 cm Hg
    B — 13.6 cm of Hg   E — 1013 cm Hg
    C — 760 cm of Hg

10. In which position do you exert the greatest pressure
    A — Standing on both feet   D — Standing on one foot
    B — Lying down   E — Standing on the tip of one foot
    C — Sitting

Questions 11-15

In the following questions you must decide if they are TRUE or FALSE. If true put T, if false put F.

11. Atmospheric pressure is caused by the weight of the atmosphere.

12. The unit of pressure is the Newton per metre.

13. When a body floats in a fluid, the density of the body must be less than or equal to the density of the fluid.

14. A hydrometer is an instrument based on the law of floatation.

15. A mercury barometer is not very accurate.

Structured Section

1. When a block of wood is placed in an eureka can, it floats and the weight of water displaced is equal to 0.4N

   a) Is the wood more dense or less dense than the water? ....

   b) What is the weight of the wood? ....

   c) Draw a diagram showing the forces which are acting on the wood when it is floating in the water.

2. The diagram is a section through part of the earth's surface.

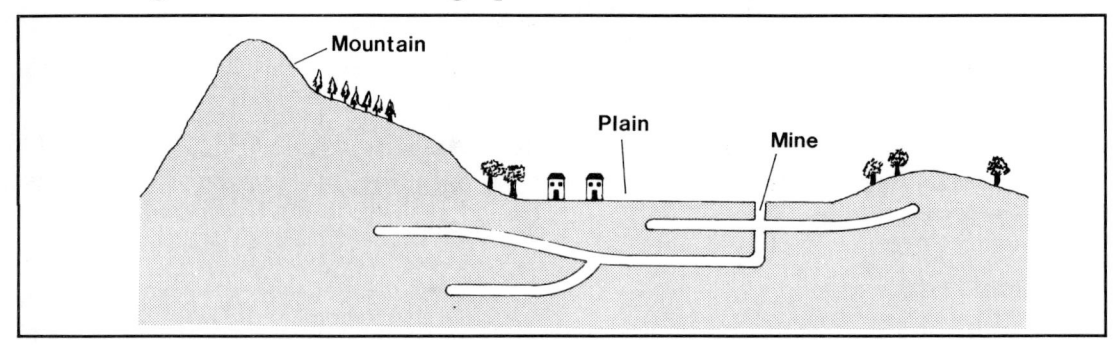

   a) On the diagram above mark the places where the maximum and minimum atmospheric pressures occur.

   b) What happens if a window breaks or a door opens on an aircraft in flight? ....
   ....

   c) What does this tell you about the pressures inside and outside the aircraft? ....
   ....

Essay Section

1. Describe an experiment you have seen which shows the large force which is exerted by atmospheric pressure. Include a diagram in your account.
2. Describe an experiment which demonstrates Archimedes' Principle.

Practical Assessment

Aim: To investigate the relationship between the depth of the depression left in plasticine, with the load placed on it.

Apparatus: Plasticine, 2 cm cube of wood, weights.

Method

1. Roll out the plasticine to an even thickness 1 cm.
2. Place wood block on top of the plasticine gently.
3. Remove the block and measure the depth of any depression.
4. Replace the block and put one of the weights on top.
5. Remove both block and weight, then measure the depth of the depression. Record the weight and depth in an appropriate table.
6. Repeat steps 1-5 using the other weights, one at a time.
7. Plot a graph of depression against weight.

34 — The Gas Laws

Multiple Choice Section

Questions 1-5

Each question is followed by five possible answers A,B,C,D and E. Choose the BEST answer.

1. The boiling point of water is:
   - A — 100°C
   - B — 37°C
   - C — 120°C
   - D — 0°C
   - E — 80°C

2. When a narrow tube is dipped into water, the water rises up the tube. This is because:
   - A — the attractive force between two water molecules has the same value as that between two glass molecules.
   - B — the attractive force between two water molecules is greater than that between a water molecule and a glass molecule.
   - C — the attractive force between a water molecule and a glass molecule is greater than that between two water molecules.
   - D — there is no attractive force between molecules of a liquid.
   - E — there is no attractive force between molecules of different substances.

3. Diffusion occurs in:
   - A — solids and liquids
   - B — gases
   - C — gases and liquids
   - D — solids and gases
   - E — solids, liquids and gases

4. Which of the following statements about the movement of molecules is correct?
   - A — move faster in a solid than in a liquid.
   - B — move slower in a gas than in a liquid.
   - C — greater in liquids than a solid.
   - D — greatest in gases and slowest in liquids
   - E — same speed in all states of matter

5. What is the value in degrees Kelvin of 60°C.
   - A — 333K
   - B — 213K
   - C — 313K
   - D — 273K
   - E — 373K

Questions 6-10

For each of the questions ONE or MORE of the answers are correct. Decide which of the responses is (are) correct. Then choose

A	B	C	D	E
1,2,3 correct	1,2 only	2,3 only	1 only	3 only

6. Which of the following are viscous liquids?
   1. syrup
   2. engine oil
   3. mercury

7. Which of the following are examples of capillary action?
   1. rising damp
   2. using a sponge
   3. sap rising in plants

8. Which of the following statements are true for Boyle's Law?
   1. the volume of the gas is inversely proportional to the pressure
   2. the temperature remains constant
   3. a fixed mass of gas is used

9. Of the following statements, which are true for Charles' Law?
   1. the gas expands by 1/273 of its volume at 0°C
   2. the temperature changes
   3. the pressure changes

10. When a substance reaches its melting point
    1. it changes from a liquid to a gas
    2. the solid and liquid state are present at the same time
    3. it changes from a solid to a liquid

Questions 11-15

In the following questions you must decide if they are TRUE or FALSE. If true put T, if false put F.

11. All substances are made up of tiny particles called atoms.

12. Liquids which flow easily have a high viscosity.

13. To convert degrees Celcius to Kelvin you add 293.

14. Boyles law state that $P \times V$ is constant.

15. Diffusion is the movement of one kind of molecule through a volume already occupied by another kind

Structured Section

1. The diagrams represent a piston and a cylinder. The trapped air cannot pass the piston.

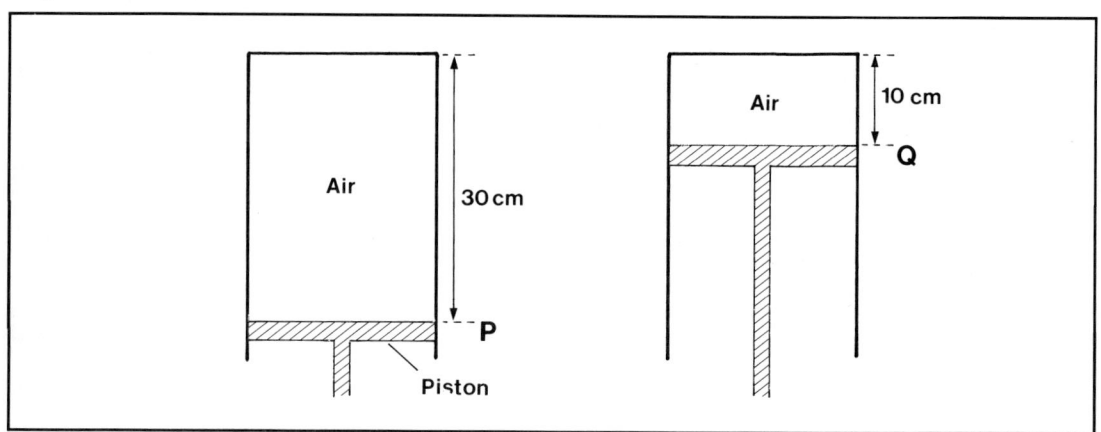

The top of the piston moves from P to Q without changing the temperature of the air.

a) What law can be used to describe how the pressure changes in the cylinder? ...

If the cross sectional area of the piston is A,

b) What is the volume of air at the beginning? ....

c) What is the final volume of the air? ....

d) If the initial pressure of the air was 70 cm of Hg, what is the final pressure .....

2. In each of the boxes below draw a diagram to represent the arrangement of the atoms in a solid, liquid and gas.

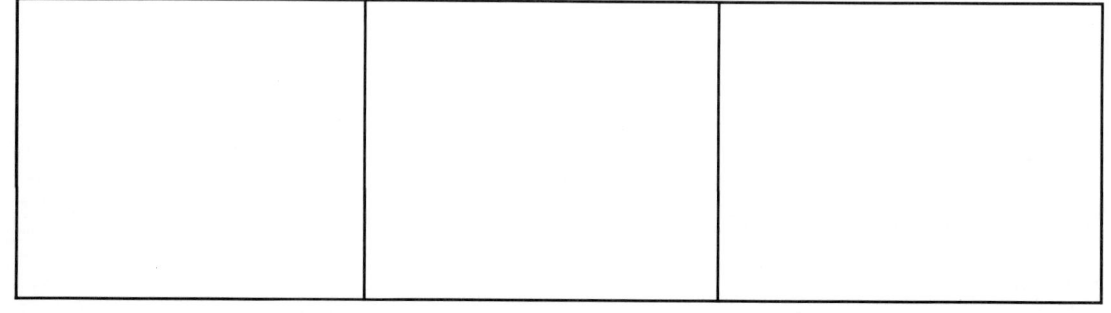

a) In which of the above states are the attractive forces between the atoms the greatest? ....

b) What effect do these strong forces have on the properties of this state of matter? ....

c) In which state are the molecules moving at a speed 1600 km/hr? ....

Essay Section

1. Describe an experiment which demonstrates that diffusion can occur in liquids. Include diagrams showing the effect of diffusion.
2. A liquid, brake fluid, is used in hydraulic brakes. Explain why a liquid is used instead of a gas and draw a diagram of the hydraulic brakes. You will need to use reference books to do this question.

Practical Assessment

Aim: To investigate the relationship between the length of a trapped mass of gas and its temperature.

Apparatus: Ice, beaker, thermometer, gauze, tripod, Bunsen burner.

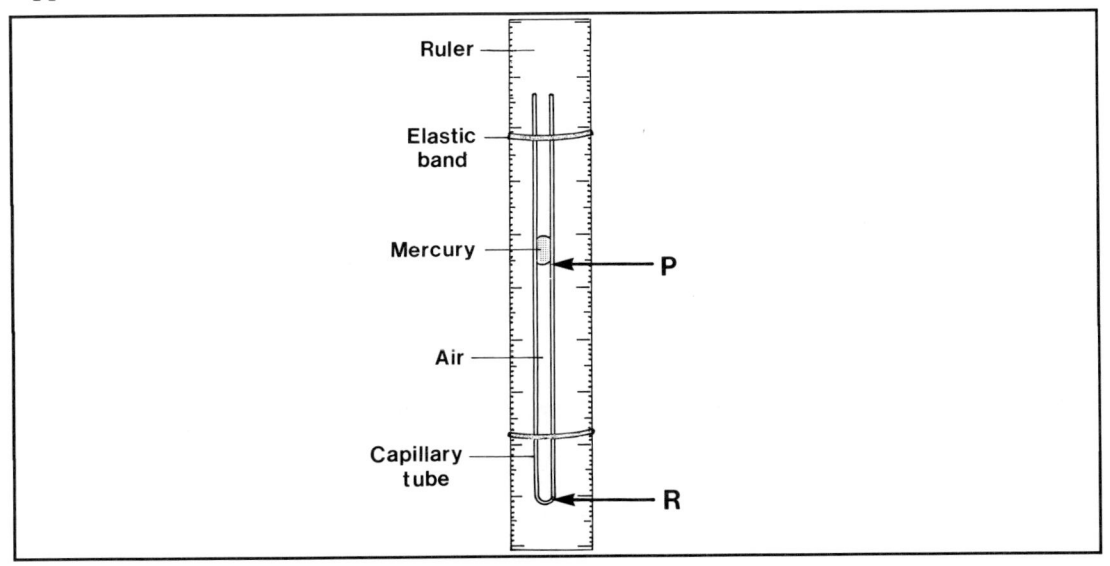

Method

1. Place the apparatus in a beaker containing enough water to cover most of the tube.
2. Take readings of the temperature of the water and the position of the bottom of the mercury index, P.
3. Now add a little ice to the water in the beaker and stir until it melts.
4. Take the temperature and note the position of P.
5. Repeat steps 3 and 4 twice more.
6. Now place the beaker on top of a tripod and heat gently for a short time.
7. Take the temperature and note the position of P.
8. Repeat steps 6 and 7 twice more and then take the reading for R.

9. Write out the results in a table, as shown below and plot a graph of length (P-R) against the temperature.

Results

Position of R =   cm.

Position of P cm	Length of air column (P-R) cm	Temperature °C

Complete Science Work-Book 2

35 — Heat Transfer

Multiple Choice Section

Questions 1-5 concern the following
 A — Conduction
 B — Radiation
 C — Immersion
 D — Convection
 E — Insulation

1. The transfer of energy in a liquid takes place by this process ☐

2. Energy travels through solids in this way. ☐

3. This piece of equipment is used to heat water in a hot water tank ☐

4. Energy reaches us from the sun by this method ☐

5. A sea breeze is an example of this type of heat transfer ☐

Questions 6-10

Each question is followed by five possible answers A,B,C,D and E. Choose the BEST answer.

6. If the reflector behind the element of an electric heater has become dull, it soon heats up when the heater is switched on. This is because:—
A — It is radiating more energy
B — It is changing more energy into light
C — It is losing less energy by convection
D — It is absorbing more energy from the element
E — Its element is conducting more energy to it

7. Energy may be transferred through a vacuum by:—
   A — Conduction only
   B — Convection only
   C — Radiation only
   D — Convection and radiation
   E — Conduction, convection and radiation

Use the following diagram to answer questions 8 and 9.

8. Which substance is the best conductor?
   A — Aluminium   D — Iron
   B — Copper   E — Wood
   C — Lead

9. Which metal is the poorest conductor?
   A — Aluminium   D — Iron
   B — Copper   E — Wood
   C — Lead

10. Fibre glass is used as insulation in the attic. It is a good insulator because:—
    A — Glass is an insulator
    B — It makes the ceiling thicker
    C — It traps lots of air between the fibres
    D — Heat passes through it
    E — It allows air to move freely

Questions 11-15

In the following questions you must decide if they are TRUE or FALSE. If true put T, if false put F.

11. In radiation, energy is 'thrown' outwards in straight lines from a hot object.

12. Dull black surfaces are poor radiators and good absorbers of radiated energy. ☐
13. Polystyrene and cotton wool are good conductors. ☐
14. A land breeze occurs when cool air from the land flows towards the sea. ☐
15. When air is heated it rises. ☐

Structured Section

1.

    Brick Air Brick   Brick

a) Cavity walls keep a house warmer in winter than solid brick walls. Why is this ....
....

b) If the cavity is filled with polystyrene beads or foam, the house is even warmer. Explain why. ....
....
....

c) Describe how a hot water radiator heats a room. ....
....
....
....

2. Supply the missing words in the following passage, from the given list.
    vibrating   atoms   solid   vibrate   hot   faster
    heated   movement   repeated   passing   moving
    conducted   conductors

The process of conduction can be explained in terms of kinetic theory. The hotter a substance is the .... its atoms .... When the end of a metal rod is ...., the atoms at the end move .... As ......

.... are very close together in a ...., the .... of the atoms at the end causes nearby .... to move faster. This process is .... along the length of the rod until all the atoms are .... In this way energy is transferred along the rod and it becomes hotter.

Essay Section

1. Describe an experiment which shows that a dull black surface is a better radiator than a bright surface. Include in your account a diagram, method and sample set of results.

2. The domestic hot water system is an example of heating by convection. Using reference books, draw a diagram showing the hot water system and explain how it works.

Practical Assessment

Aim: To investigate the insulation properties of various materials

Apparatus: Imitation fur (long and short fibre), cotton wool, feathers, imitation lambswool, cans, beakers, flasks, thermometers, bench mats, supply of hot water and ice.

Instructions

1. Using some or all of the equipment provided, design an experiment to test how good cotton wool, feathers, imitation fur and wool are as insulators.

2. Discuss with your teacher whether your ideas will work.

3. Collect the apparatus you require and then do the experiment.

4. Write out an account under the headings: Apparatus, Method, Results and conclusions.

36 — Reflection

Multiple Choice Section

Questions 1-5 concern the following diagram

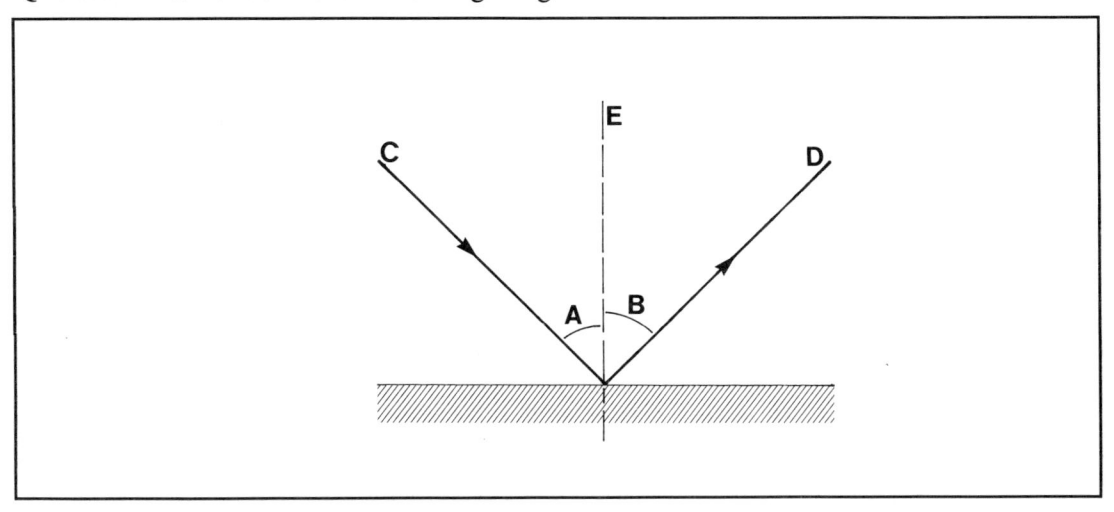

On the above diagram which letter marks
1. The reflected ray? ....
2. The angle of incidence? ....
3. The normal? ....
4. The incident ray? ....
5. The angle which must be 0° for the light to return along the same path. ....

Questions 6-10

For each of the question ONE or MORE of the answers are correct. Decide which of the responses is (are) correct. Then choose:

A	B	C	D	E
1,2,3 correct	2,3 only	2,3 only	1 only	3 only

6. Which of the following is/are a type of image?
   1. Inverted
   2. Real
   3. Virtual

7. In regular reflection
   1. The reflecting surface is smooth
   2. The incident and reflected rays are in the same plane as the normal

3. The angle of incidence is equal to the angle of reflection ☐

8. The image formed by a plane mirror is
   1. Virtual
   2. Laterally inverted ☐
   3. Larger than the object

9. Multiple images can be formed by

1   2   3

10. ☐

Using the diagram above, which of the following are true for a concave mirror?
1. The rays of light diverge after reflection
2. Another name is the converging mirror ☐
3. It has a real focus

Questions 11-15

In the following questions you must decide if they are TRUE or FALSE. If true put T, if false put F.

11. A convex mirror is known as a converging mirror. ☐

12. A concave mirror can produce a magnified but virtual image. ☐

13. Parallax is when two fixed objects appear to move relatively to each other. ☐

14. Diffuse reflection occurs when the surface is rough. ☐

15. The image produced by a plane mirror is twice the distance behind the mirror as the object is in front. ☐

Structured Section

1.

a) On the diagram above draw in the paths of the three reflected rays of light

b) Mark on the diagram the focus and the focal length of the mirror.

c) Draw in the space below a similar diagram for a convex mirror, indicating the position of the focus and focal length.

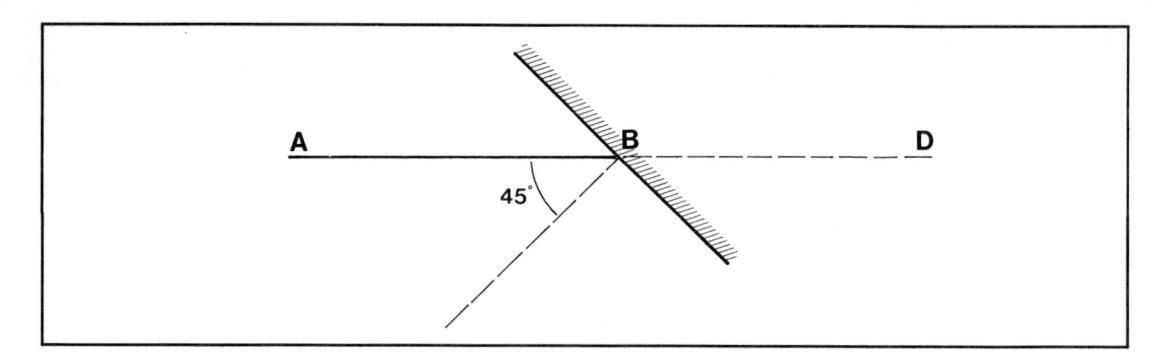

2. The diagram above shows a ray of light striking a plane mirror with an angle of incidence of 45°.

   a) Draw accurately, using a protractor, the reflected ray and label it BC.

   b) Measure the angle CBD. This is the angle through which the incident ray has been turned ....

   c) Draw a diagram of an instrument used to look over the heads of a crowd in the space below.

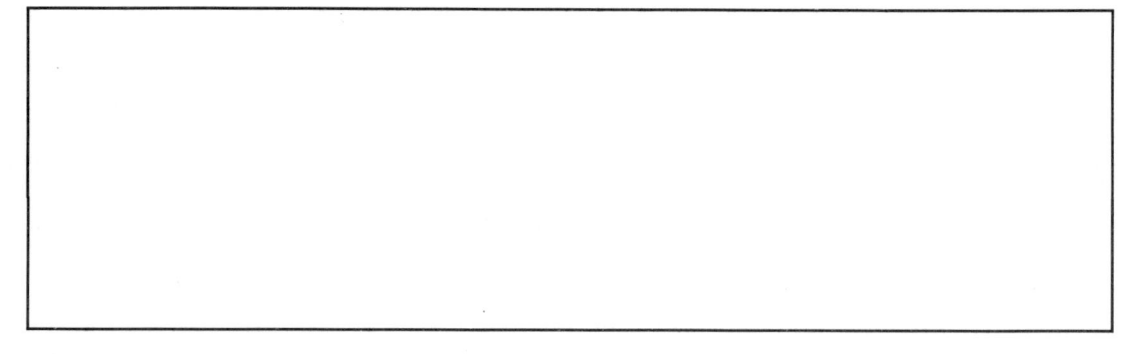

Essay Section

1. It is stated that the image produced by a plane mirror is as far behind the mirror as the object is in front. Describe an experiment you have carried out which proves that this is correct.

2. Describe with the aid of a diagram, the construction of a Kaleidoscope and the pattern observed.

Practical Assessment

Aim: To investigate the relationship between the incident angle and the deviation of the incident ray.

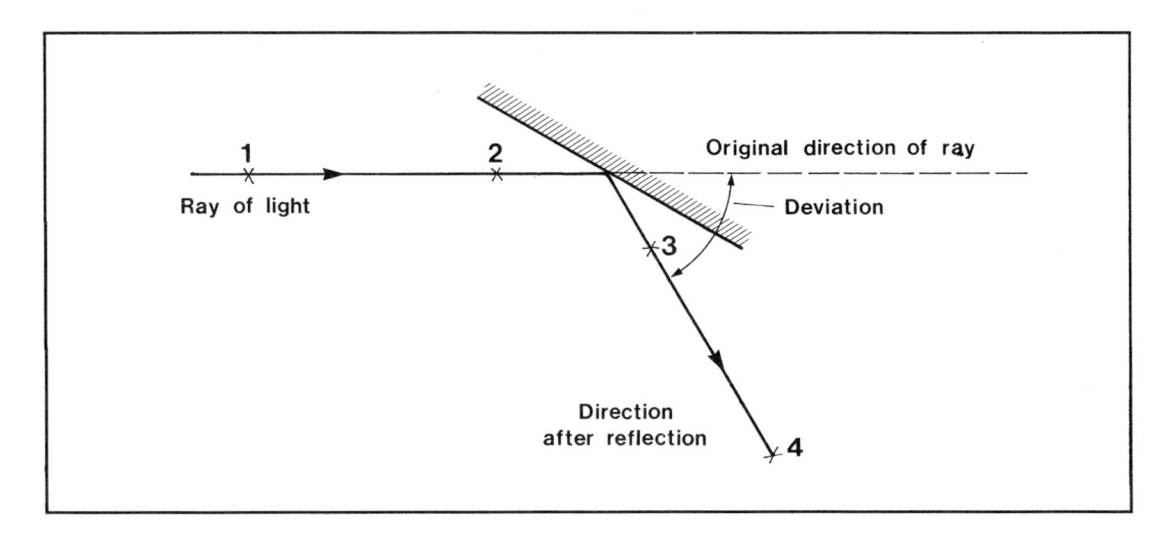

Information

When a ray of light strikes a mirror, it is deviated. This means it does not continue in the same direction. The angle between the original direction of the ray and the reflected ray is the angle of deviation of the ray.

Method

1. Set up a mirror on edge on a plain piece of paper and draw a line along the back. Direct a ray of light towards the mirror.
2. Using a pencil, mark its position with crosses 1 and 2.
3. Also mark the position of the reflected ray with crosses 3 and 4.
4. Remove all apparatus and draw a solid line through points 1 and 2 until it cuts the mirror surface. Then continue as a dotted line.
5. Join up points 4 and 3, and extend to touch the mirror line.
6. Measure the angle between the dotted line and the reflected ray. Record your results in the table below.
7. Replace all apparatus, change the angle at which the light ray strikes the mirror and then repeat steps 2-6.
8. Repeat step 7 another **four** times.
9. Write a short report, which includes the table and a conclusion.

Angle of Incidence	Deviation

Complete Science Work-Book 2

37 — Refraction

Multiple Choice Section

Questions 1-5

Each question is followed by five possible answers A,B,C,D and E. Choose the BEST answer.

1. Which of the following diagrams is correct?

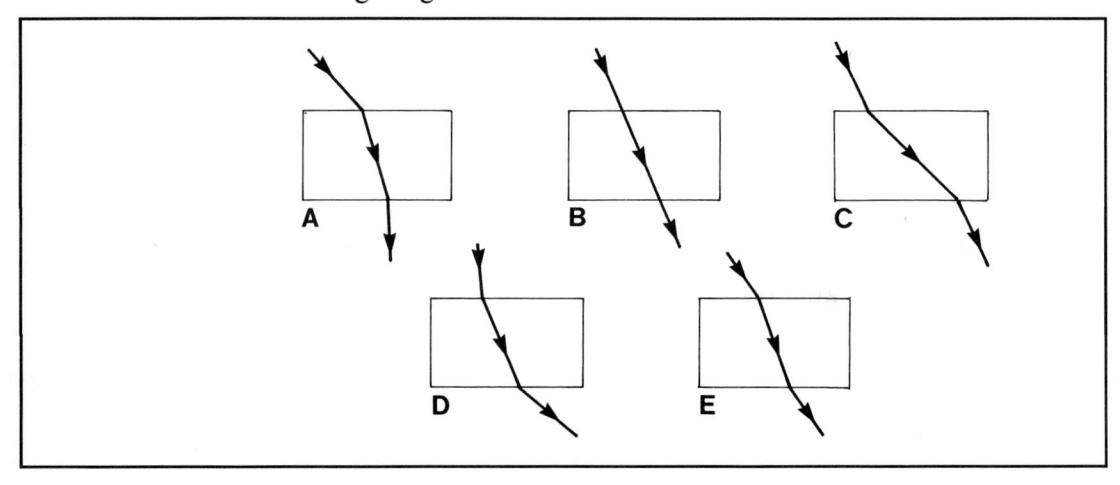

2. The refractive index, n is given by:

   A — $\dfrac{\text{Sin r}}{\text{Sin i}}$

   B — Sin i × Sin r

   C — Sin i

   D — $\dfrac{\text{Sin i}}{\text{Sin r}}$

   E — Sin r

3.

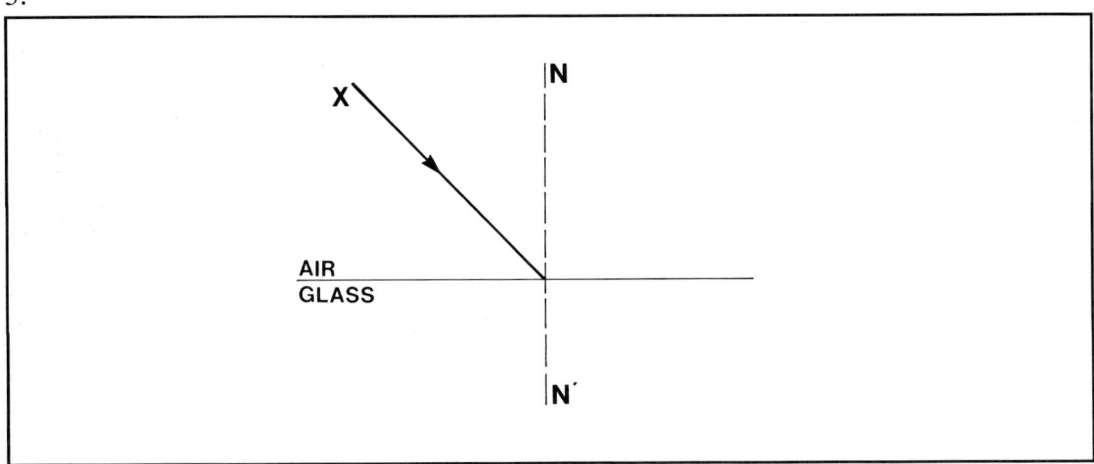

The line marked X on the diagram is the:
- A — Refracted ray
- B — Normal
- C — Surface
- D — Medium
- E — Incident ray

4. Total internal reflection can occur when a light ray:
- A — Skims the surface of a block of glass
- B — Strikes the surface of the glass block at right angles
- C — Moves from a denser medium into a less dense medium
- D — Passes from a less dense medium into a denser one
- E — Has an incident angle of 0° at the air/glass surface

5. Which of the following lenses is NOT a convex lens

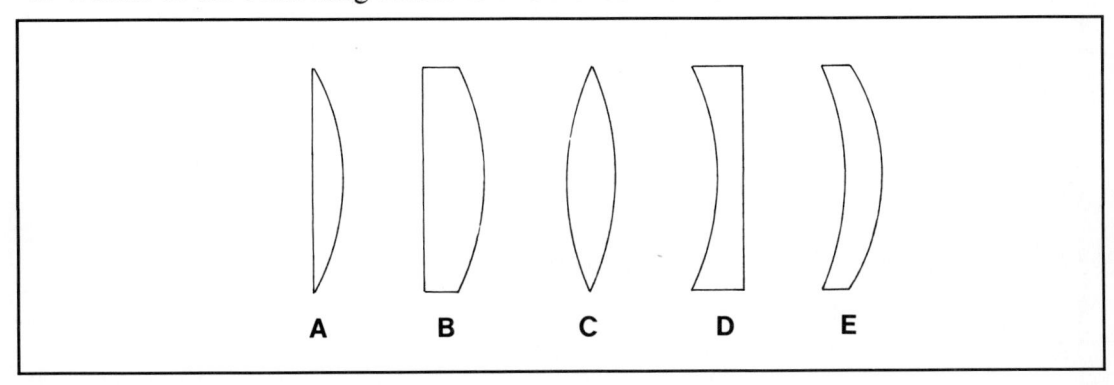

Questions 6-10

For each of the questions ONE or MORE of the answers are correct. Decide which of the responses is (are) correct. Then choose:

A	B	C	D	E
1,2,3 correct	1,2 only	2,3 only	1 only	3 only

6. Which of the following statements describe the way in which light is refracted?
   1. The incident and refracted rays are on opposite sides of the normal at the point of incidence.
   2. The ratio of the sine of the angle of incidence to the sine of the angle of refraction is constant.
   3. The incident ray, refracted ray and normal all lie in the same plane ☐

7. Which of the following statements are correct?
   1. The angle of incidence equals the angle of refraction.
   2. The normal is a line drawn at 90° to the surface where the incident ray strikes it.
   3. The refractive index of glass is 1.5 ☐

8. A concave lens:
   1. Is known as a diverging lens
   2. Is thinner at the centre than at the edges
   3. Brings rays of light together after they have passed through it ☐

9. A mirage is formed because:
   1. The air is hotter near the ground
   2. The air is less dense near the ground
   3. At one point the light rays are totally internally reflected. ☐

10. A 45° prism can turn a light ray through:
    1. 45°
    2. 90°
    3. 180° ☐

Questions 11-15

In the following questions you must decide if they are TRUE or FALSE. If true put T, if false put F.

11. The angle between the normal and the incident ray is called the angle of incidence. ☐

12. When a ray of light strikes a surface at right angles the angle of incidence is 90°. ☐

13. A fibrescope is a thin flexible light pipe. ☐

14. In prism binoculars 45° prisms are used to turn light rays through 180°. ☐

15. The distance between a convex lens and the point where initially parallel rays of light are brought together is called the focus of the lens. ☐

Structured Section

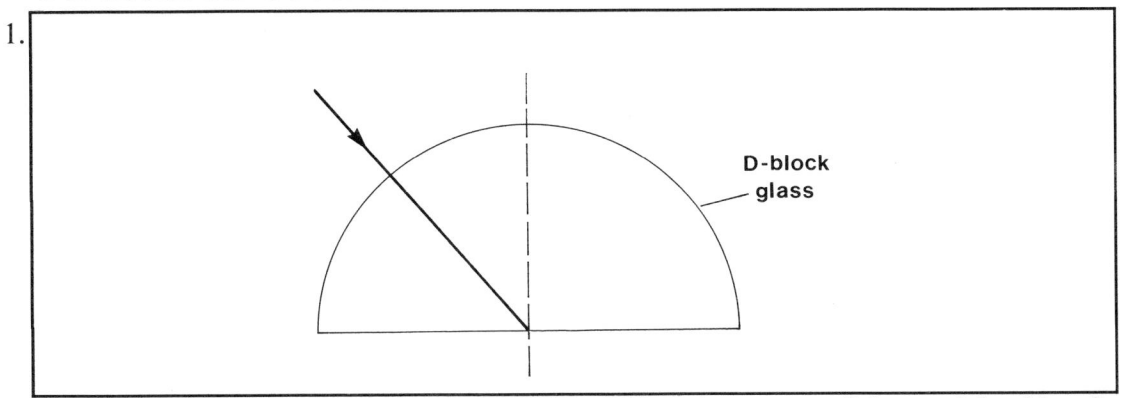

1.

The diagram above shows a ray of light striking the centre of the straight side of a D-shaped block of glass at the critical angle.

On the diagram
a) Mark the critical angle using the letter C
b) Complete the path of the ray out of the block
c) Draw in another ray to show total internal reflection

2. On the diagram of the periscope below, draw in two light rays from the object to the observer.

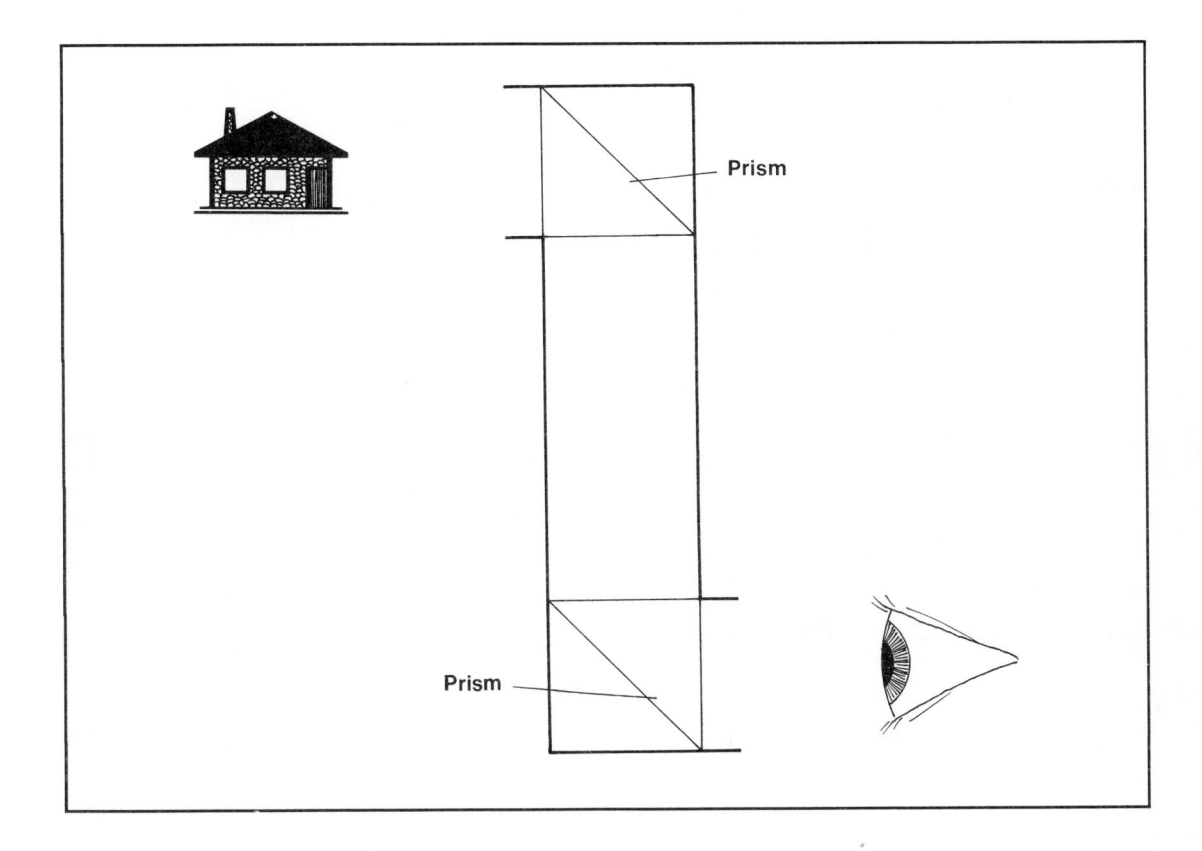

a) Describe what happens as a ray passes through one of the prisms ....
   ....
   ....

b) What, if anything, will happen to the image seen if the length of the periscope is increased? ....
   ....
   ....

Essay Section

1. Describe the way in which 45° prisms are used in binoculars and draw a diagram to illustrate your answer.
2. Describe an experiment which you have carried out that shows the effect of a convex and a concave lens on a parallel beam of light.

Practical Assessment

Aim: To investigate the effect on lateral displacement of changing the thickness of the glass block.

Apparatus: Ray box, pins, protractor, soft board, plain and ruled paper, microscope slides, rubber bands, plasticine, pencil, ruler.

Method

1. Design an experiment to study the effect of changing the thickness, d, of the glass on the lateral displacement.

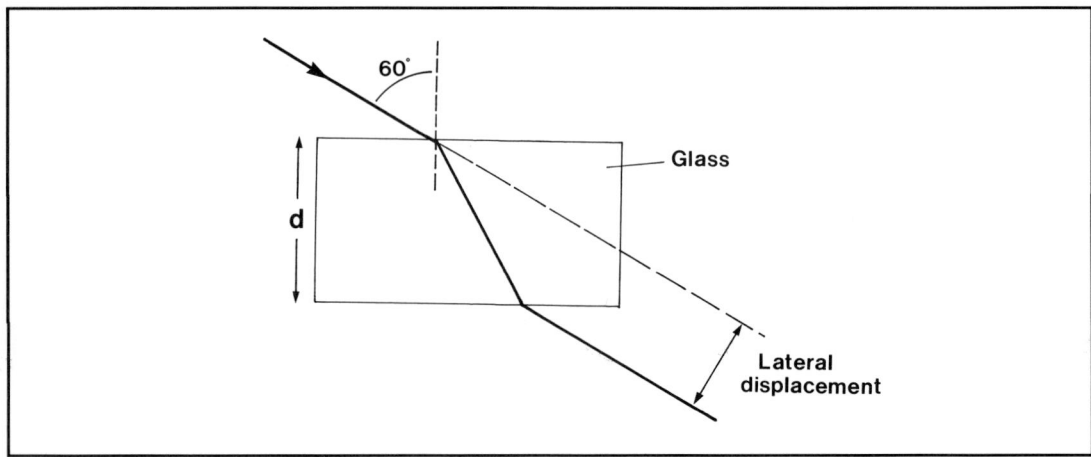

2. Write out your method on the ruled paper and discuss your ideas with your teacher.
3. Set up the apparatus and using a 60° incident angle throughout, carry out the experiment.
4. Set out the results in an appropriate table.

1. Microscope slides can be used to produce a glass block of variable thickness, d. They are arranged as shown below.

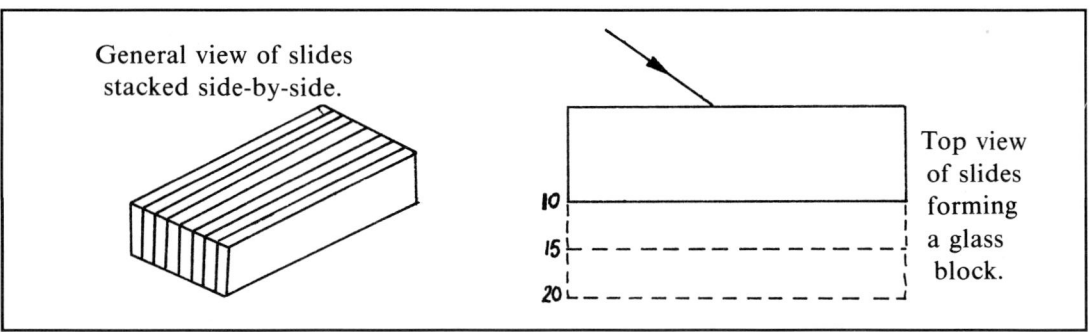

General view of slides stacked side-by-side.

Top view of slides forming a glass block.

2. Starting with 5 slides, set up the apparatus as shown in the diagram using a 60° incident angle.
3. Measure the lateral displacement which occurs.
4. Repeat steps 2 and 3 using 10, 15, 20, 25 and 30 slides.
5. Set out the results in an appropriate table and plot a graph of lateral displacement against thickness of the glass blocks.
6. What happens to the lateral displacement as the block becomes thicker?

Complete Science Work-Book 2

38 — Optical Instruments

Multiple Choice Section

Questions 1-5

Each question is followed by five possible answers A,B,C,D and E. Choose the BEST answer.

1. When a magnifying lens is held close to an object, it produces an image which is:—
   A — Real and magnified
   B — Inverted, real and magnified
   C — Virtual and inverted
   D — Magnified and upright
   E — Diminished, virtual and inverted

2. A camera is focused by:—
   A — Opening the shutter
   B — Changing the size of the aperture
   C — Changing the shape of the lens
   D — Altering the distance between the lens and the film
   E — Winding the film on

3. When light enters the eye, most of the bending occurs at the
   A — Retina   D — Iris
   B — Lens   E — Pupil
   C — Cornea

4. Which of the following statements is INCORRECT?
   A — A camera takes one picture at a time, the eye continually scans the scene
   B — In very bright light the pupil of the eye becomes very small
   C — The near point for normal vision is 25 cm from the eye
   D — A reel of cine film consists of thousands of still pictures
   E — The rods in the retina are sensitive to colour

5. Which of the following lenses could not be used to correct long-sight?

150

Questions 6-10

For each of the questions ONE or MORE of the answers are correct. Decide which of the responses is (are) correct. Then choose:

A	B	C	D	E
1,2,3 correct	1,2 only	2,3 only	1 only	3 only

6. The amount of light entering a camera is controlled by
   1. Adjustable aperture
   2. Shutter
   3. The distance between the lens and the film

7. Both the camera and the eye contain this/these
   1. A light sensitive screen
   2. An iris
   3. Variable aperture

8. The image formed on the retina is
   1. Virtual
   2. Real
   3. Inverted

9. A person who suffers from short-sight possibly has
   1. An eyeball which is too long
   2. A lens which is too curved
   3. A retina which is damaged

10. To which of the following colours are the cones in the eye sensitive?
    1. Yellow
    2. Red
    3. Blue

Questions 11-15

In the following questions you must decide if they are TRUE or FALSE. If true put T, if false put F.

11. Rods and cones are light sensitive cells in the retina.
12. A concave lens can be used as a magnifying glass.
13. The eye and the camera are light-tight boxes.
14. A person who can see near objects clearly but cannot focus distant objects is long sighted.
15. If you only have one eye, it is difficult to judge distance.

Structured Section

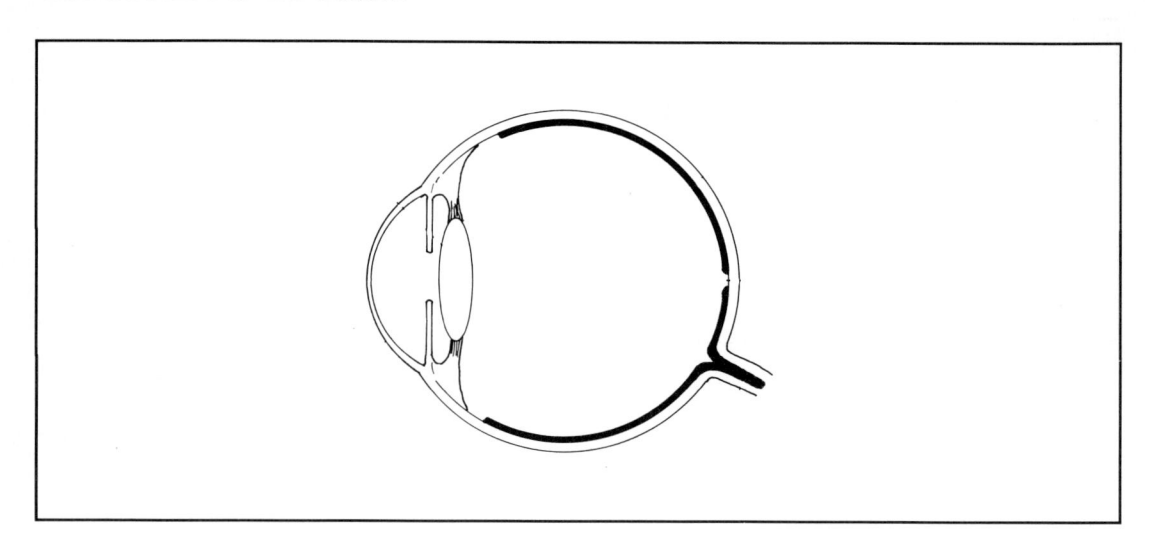

1. a) On the above diagram, label the lens, pupil, retina, and optic nerve.

   b) Describe the function of the following:

   Retina ....

   ....

   Pupil..

   ....

   Optic Nerve...

   ....

   c) If the eye is short-sighted a lens must be placed in front to correct this defect. Draw in the space below a diagram to show this correction.

2.

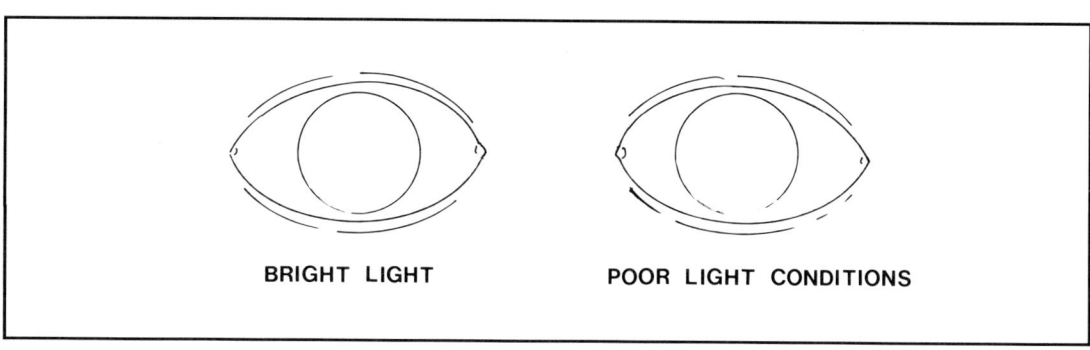

BRIGHT LIGHT   POOR LIGHT CONDITIONS

a) Complete the diagrams above to show the effect of changing light conditions, on the pupil.

b) Why does the size of the pupil change? ....
....
....

c) What is the effect of too much light reaching the retina? ....
....
....

Essay Section

1. Describe an experiment you have carried out that shows it is very important for predatory animals to have both eyes directed forwards.
2. Draw a labelled diagram of a camera. Explain how it is focussed and why a shutter is needed.

Practical Assessment

Aim: To make a simple telescope.

Apparatus: Metre rule, 5 cm and 20 cm lens, plasticine.

Information

Another optical instrument is a telescope. This instrument enables us to see distant objects more clearly.

Method

1. Collect a metre rule, a 5 cm lens, a 20 cm lens and some plasticine.

2. Then securely attach the 5 cm lens (eyepiece) to the ruler at the 10 cm mark using the plasticine.

3. Place the 20 cm lens at distance of 25 cm away from the eyepiece. Again use plasticine to attach it securely to the ruler.

4. Lift up the rule (take care not to drop the lenses) and point it towards the window. Look through the eyepiece.

5. Can you see an image? If not move the 20 cm lens until a sharp image is seen.

6. Answer the following questions:
   a) Is the image magnified?.....................................
   b) Is the image upside down or upright? ....

Complete Science Work-Book 2

39 — Waves and Sound

Multiple Choice Section

Questions 1-5 concern the following
- A — Transverse wave
- B — Wavelength
- C — Amplitude
- D — Longitudinal wave
- E — Frequency

1. Ripples are example of this type of wave

2. This is the number of waves passing a particular point in one second

3. Compressions and rarefactions occur in ths type of wave

4. This is the distance between one crest and the next

5. The velocity of a wave can be found when wavelength is multiplied by

Questions 6-10

Each question is followed by five possible answers A,B,C, D and E. Choose the BEST answer.

6. When a water wave moves from one point to another
   - A — The water moves between the points
   - B — the water does not move
   - C — Energy is carried from one point to the other
   - D — There is no movement of energy
   - E — Both the water and energy move between the points

7.

In the wave diagram, which distance A, B, C, D or E is the amplitude?

155

8. The note produced by a sonometer depends on the following
   A — Length and tension
   B — Length
   C — Thickness and tension
   D — Tension
   E — Length, tension and thickness of the wire

9. Which one of the following is an example of LONGITUDINAL waves?
   A — Ripples on water
   B — Microwaves
   C — Soundwaves
   D — Radio waves
   E — Waves on a rope

10.

If the pendulum P is made to swing, soon Q also starts to swing with the same frequency. After a while P stops and then the process is reversed. This effect is known as:
   A — Vibration
   B — Resonance
   C — Radiation
   D — Oscillation
   E — Rarefaction

Questions 11-15

In the following questions you must decide if they are TRUE or FALSE. If true put T, if false put F.

11. In a transverse wave the medium vibrates in the same direction as the energy is travelling.

12. Sound is a form of energy. All sounds come from vibrating objects.

13. The pitch of a note depends on the frequency.

14. Sound cannot travel through a vacuum.

15. An open organ pipe has the same frequency as a closed pipe of the same length.

Structured Section

1.

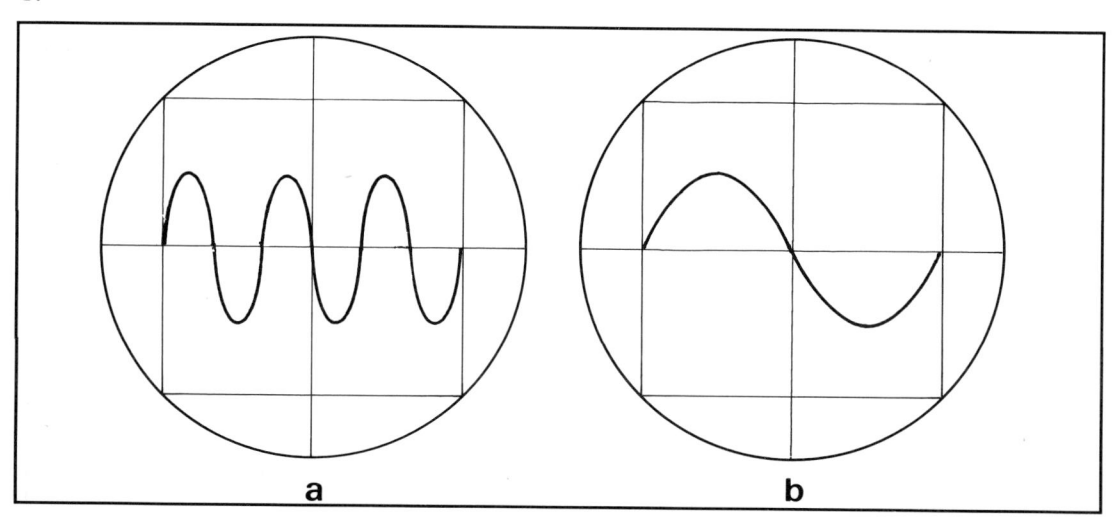

A microphone is connected to a cathode ray oscilloscope. When a vibrating tuning fork of frequency 512Hz is held close to it, trace (a) is produced on the screen.

a) A second fork produces trace (b). Is the frequency of this fork greater or less than 512Hz? ....

b) In the space below copy (b) and label the wavelength and the amplitude of the trace.

c) How does the wavelength change from (a) to (b)? ....
....

d) If the tuning fork in (a) was moved a short distance away from the microphone how would the trace change? Draw a diagram to illustrate your answer.

2. A fishing boat using sonar to detect shoals of fish sends down a short pulse of soundwaves. The echo is received 1/10s later. If the speed of sound in water is 1500m/s:

   a) How far has the pulse travelled in 1/10s? ....

   b) How far below the boat is the shoal of fish? ....

   c) The reflected pulse lasts longer than the emitted pulse ....

   Suggest a reason for this ....

   ....

Essay Section

1. Discuss with the use of diagrams the similarities and differences between transverse and longitudinal waves.
2. Describe in detail, an experiment to determine the factors which affect the note emitted by a stretched string.

Practical Assessment

Aim: To investigate the reflection of sound waves

Apparatus: 2 cardboard tubes, sheet of glass or hard smooth surface, plasticine to support glass and screen, screen, watch or clock with loud tick, sheet of paper.

Method

1. Set up apparatus as shown below, on the sheet of paper.

Complete Science Work-Book 2

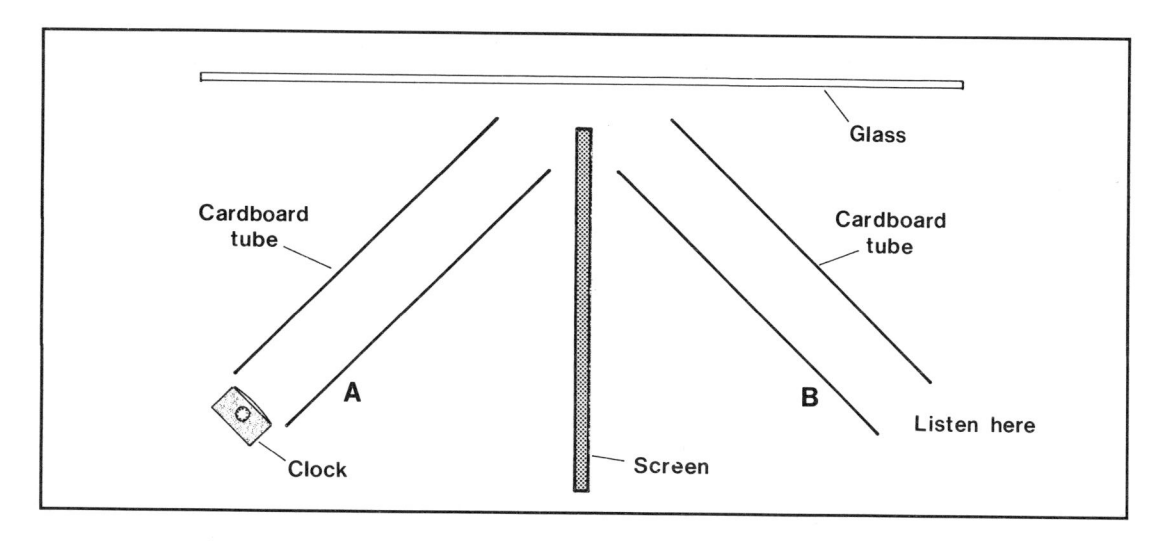

2. Move tube B around until the ticking of the watch is heard quite clearly.
3. Mark the position of the tubes A and B by drawing along the sides, and also that of the glass.
4. Remove all apparatus from the paper and extend the lines from A and B to the glass. If they do not initially meet draw a line parallel to one of them so you have a diagram as shown below.

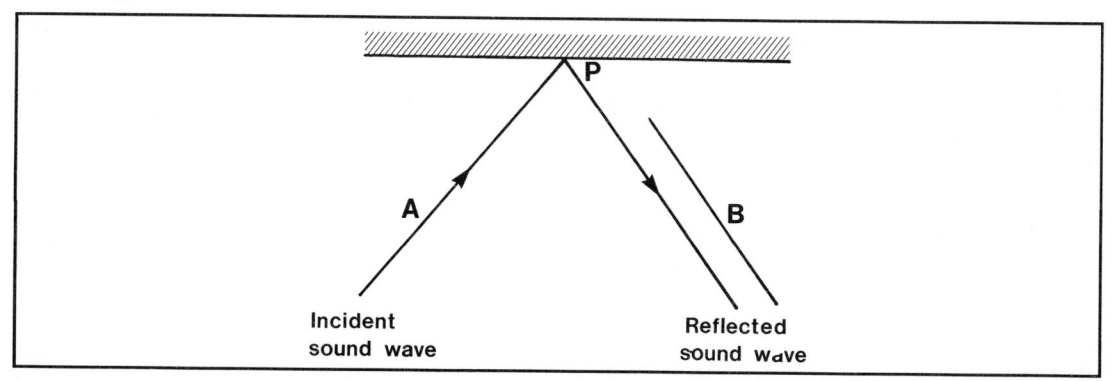

5. Draw in the normal at point P and measure the angles of incidence and reflecion.
6. Repeat the experiment for different incident angles.
7. Write up your results and decide whether sound waves obey the laws of reflection which are:—
   a) the angle of reflection is equal to the angle of incidence
   b) the incident ray, reflected ray and normal all lie in the same plane.

Complete Science Work-Book 2

40 — Static Electricity

Multiple Choice Section

Questions 1-15 concern the following

    A — Charged
    B — Conductor
    C — Electrons
    D — Insulator
    E — Induction

1. Plastic is an

2. These are negatively charged

3. The leaves of an electroscope move apart when it is

4. When perspex is rubbed with a cloth, these are transferred to the cloth

5. Substances which allow charge to flow through them

Questions 6 - 10

For each of the questions ONE or MORE of the answers are correct. Decide which of the responses is (are) correct. Then choose

A	B	C	D	E
1,2,3 correct	1,2 only	2,3 only	1 only	3 only

6. If a polythene rod is rubbed using a cloth
    1. The polythene has a negative charge
    2. The cloth has an equal but opposite charge to the polythene
    3. Neither of them becomes charged

7. When a negative charged rod is held close to an electroscope
    1. The leaves diverge
    2. The charge on the electroscope is positive
    3. The rod loses its charge to the electroscope

8. Which of the following are found in the nucleus of an atom?
    1. Electrons
    2. Neutrons
    3. Protons

9. An electroscope can be used to test
    1. If a body is an insulator or conductor
    2. If a body is charged
    3. The sign of a charge

10.

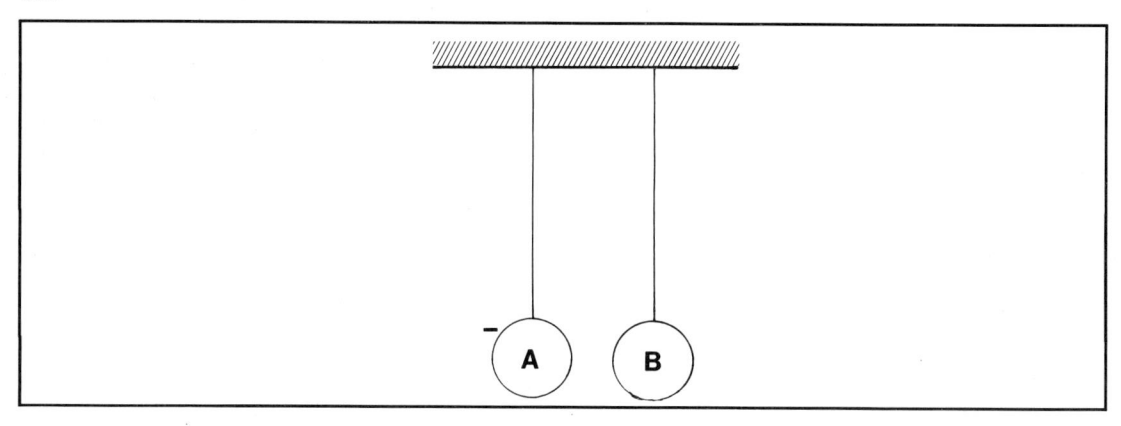

If the sphere B is given a negative charge it
1. Moves towards A
2. Remains stationary
3. Moves away from A

Questions 11-15

In the following questions you must decide if they are TRUE or FALSE. If true put T, if false put F.

11. There are two types of charge, positive and negative.

12. Lightning is an example of static electricity.

13. Like charges attract one another, unlike charges repel each other.

14. A Van de Graaf generator produces very small static charges.

15. When an electroscope is charged by induction, the cap is touched with a finger. This is called earthing the electroscope.

Structured Section

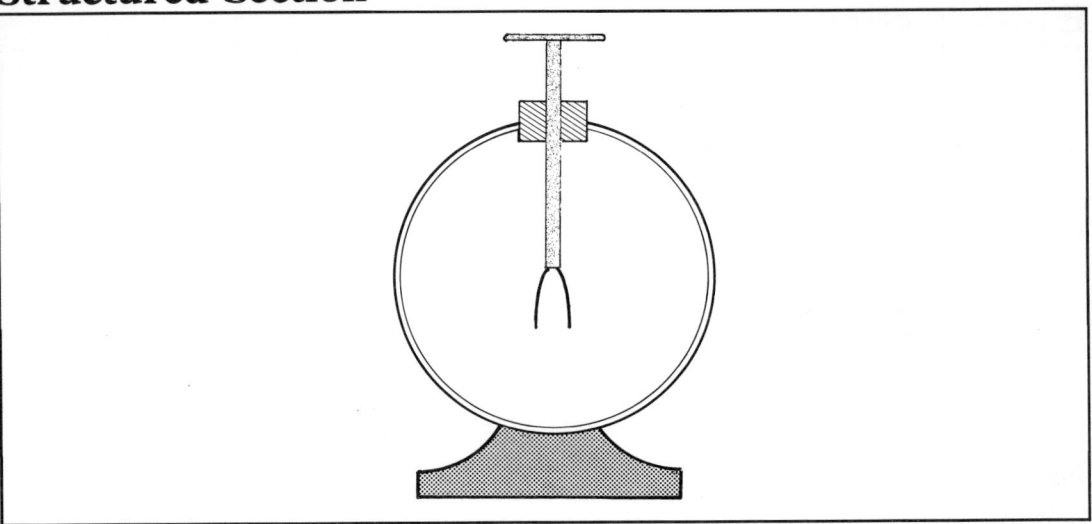

1. a) On the diagram label the cap, insulator and leaves.

   b) Describe what happens to the leaves when a positively charged rod is brought near the cap...
   ....
   ....

   c) Draw in the positively charged rod and show the position of the charges on the electroscope.

2. Fill in the missing words in the following paragraph. Choose the appropriate words from the list below.

   leaves,   earthing,   negative,   cap,   rubbing,   apart,
   positive,   together,   ground,   case,   rod)

   In order to charge an electroscope by induction the following method is used.

   Charge a polythene rod by .... and hold it near the .... of the electroscope. The result of this step is that the .... charges in the electroscope are repelled by the negative charges on the rod. The negative charges thus gather on the .... leaving the .... with a positive charge. As a result of the .... charges on the leaves they move ....
   .....
   Touch the .... with your finger. This allows the negative charges on the .... to flow to the .... through your body and is called . ... the electroscope.

Essay Section

1. Describe the experiment used to show the forces which exist between charged bodies. Include a diagram in your account.
2. Using the text book and other reference books write an essay on 'Thunder and Lightning'.

Practical Assessment

Aim: Devise an experiment to investigate the charges produced when:
   glass is rubbed with silk
   ebonite is rubbed with fur

Apparatus: glass rod, ebonite rod, silk, fur, perspex and polythene rods, electroscope.
   Other apparatus available on request.

Instructions

1. Decide how you will carry out this experiment and then check that your method will work with your teacher.
2. Collect your apparatus and do the experiment.
3. Write a report which includes a list of apparatus, method and results. Use the results to decide which charges (negative or positive) are found on ebonite and glass rods after rubbing.

41 — Current Electricity

Multiple Choice Section

Questions 1-5

Each question is followed by five possible answers, A,B,C,D and E. Choose the BEST answer.

1. The charge carriers in metals are:
   - A — Neutrons
   - B — Protons and Neutrons
   - C — Protons
   - D — Electrons
   - E — Protons and Electrons

2. Which of the following is the unit of electric current?
   - A — Volt
   - B — Ohm
   - C — Watt
   - D — Ampere
   - E — Coulomb

3. In the simplest type of cell, the chemical reactions cannot be reversed. This type of cell is called:
   - A — Accumulator
   - B — Simple cell
   - C — Primary cell
   - D — Strorage cell
   - E — Secondary cell

4. What does the following symbol represent?

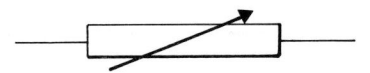

   - A — A fixed resistor
   - B — A fuse
   - C — A switch
   - D — A lamp
   - E — A variable resistor

5. What is the total voltage of the arrangement shown below if each cell has a voltage (potential difference) of 1.5 v

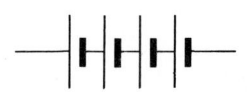

   - A — 1.5 v
   - B — 6.0 v
   - C — 3.0 v
   - D — 0v
   - E — 4.5 v

Questions 6-10

For each of the questions ONE or MORE of the answers are correct. Decide which of the responses is (are) correct. Then choose:

A	B	C	D	E
1,2,3 correct	1,2 only	2,3 only	1 only	3 only

6. Potential difference
   1. Is measured in volts
   2. Called a voltage
   3. Can be called an electrical pressure

7. Which of the following can be found in a simple cell?
   1. Electrodes
   2. Electrolyte
   3. Dilute sulphuric acid

8. In which circuit/circuits would the lamp light?

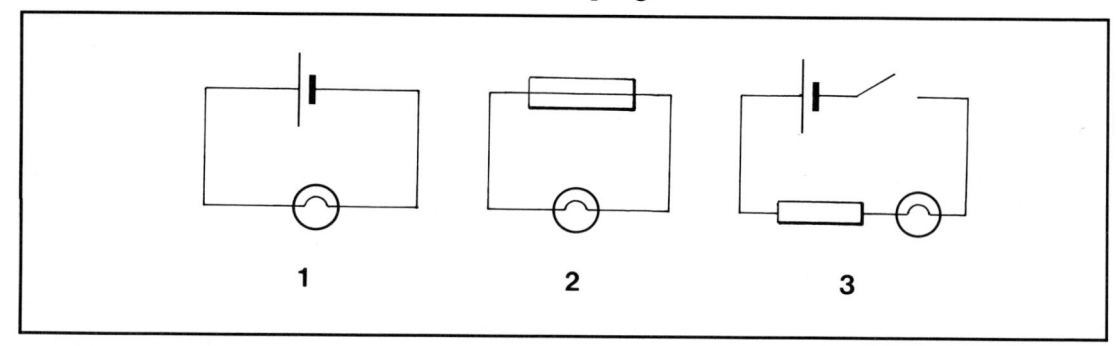

9. Which symbol/symbols represent a fuse?

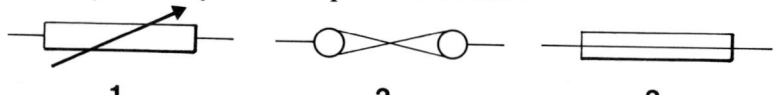

10. The value of the electrical resistance of a material is due to:
    1. The force between the moving electrons and stationary atoms
    2. The number of electrons which can move
    3. The total number of electrons in the material

Questions 11-15

In the following questions you must decide if they are TRUE or FALSE. If they are true put T, if false put F.

11. Electric current is the flow of charge from one place to another.

12. Voltage = Current/Resistance.

13. The resistance of a conductor is a measure of its opposition to the movement of charge through it. ☐

14. Variable resistors are called rheostats. ☐

15. The current supplied by the Electricity Board is direct current. ☐

Structured Section

1. a) Complete the symbol table:

SYMBOL	MEANING
—❘⊦—	CELL.
———	
—ww—	
—╱ —	
—(A)—	
⏚	

b) Using symbols, draw circuits containing the following components in the space below.

(i) a cell, an ammeter and a lamp
(ii) a battery, switch and two resistors in series
(iii) a battery, switch, ammeter, resistance and voltmeter measuring the voltage across the resistance.

2.

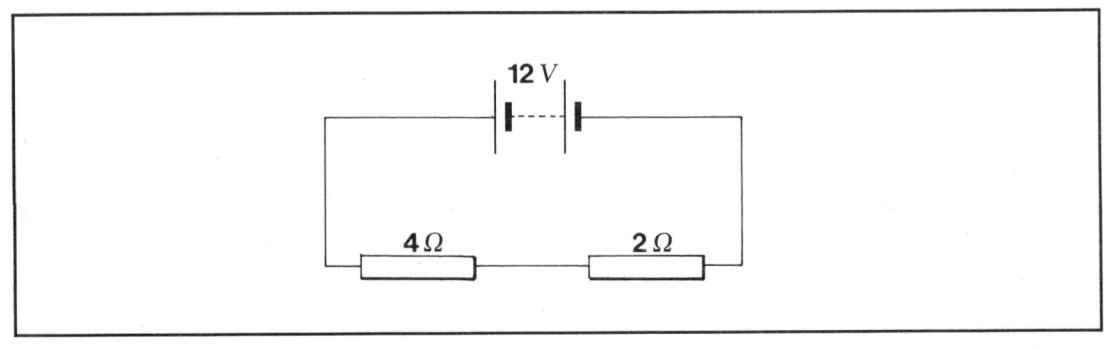

a) What is the battery voltage in the above circuit? ....

b) If the battery is made from zinc-carbon cells; how many cells does it contain? ....

c) What is the voltage across the two resistors? ....

d) What is the total resistance in the circuit? ....
e) What is the current flowing through the circuit? ....
f) What is the potential difference across the 4Ω resistors? ....

Essay Section

1. The size of a current flowing in a circuit is said to depend on the voltage applied to the circuit and also the resistance of the circuit. Describe the two experiments you have performed which demonstrate that this statement is true.
2. Draw a labelled diagram of a dry cell and describe its construction.

Practical Assessment

Aim: To find the resistance of a piece of wire.

The following results were obtained in an experiment to measure the resistance of a piece of wire.

V/volts	I/amps	V/volts	I/amps
1	0.2	6	1.2
2	0.4	7	1.4
3	0.6	8	1.6
4	0.8	9	1.8
5	1.0	10	2.0

Instructions

(a) Plot the results on the graph below, choose suitable scales for each axis.

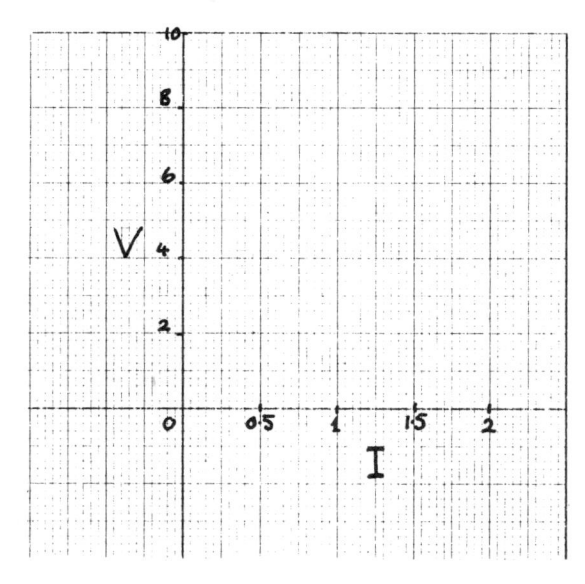

(b) From the graph or otherwise, find the resistance of the wire.

42 — Effects of Electric Current

Multiple Choice Section

Questions 1-5

Each question is followed by five possible answers A,B,C, D and E. Choose the BEST answer.

1. The unit of power is
   - A — Ohm
   - B — Watt
   - C — Joule
   - D — Volt
   - E — Ampere

2. When a voltage V is applied to an electric fire element a current I flows. The power of the fire is:
   - A — I^2V
   - B — V/I
   - C — IV
   - D — I/V
   - E — I^2/V

3. When a current flows through a wire it produces a magnetic field. The direction of the magnetic field lines are:

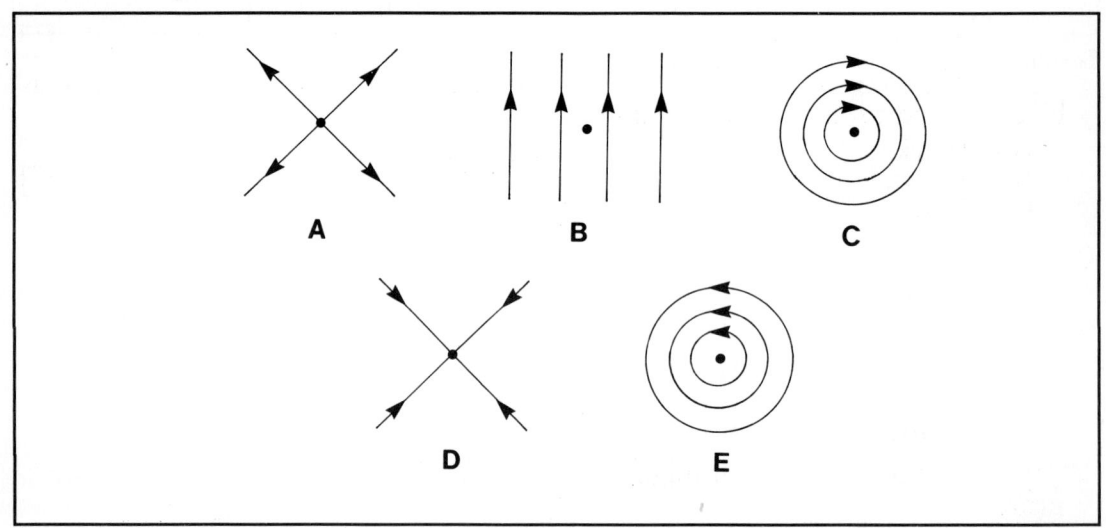

4.

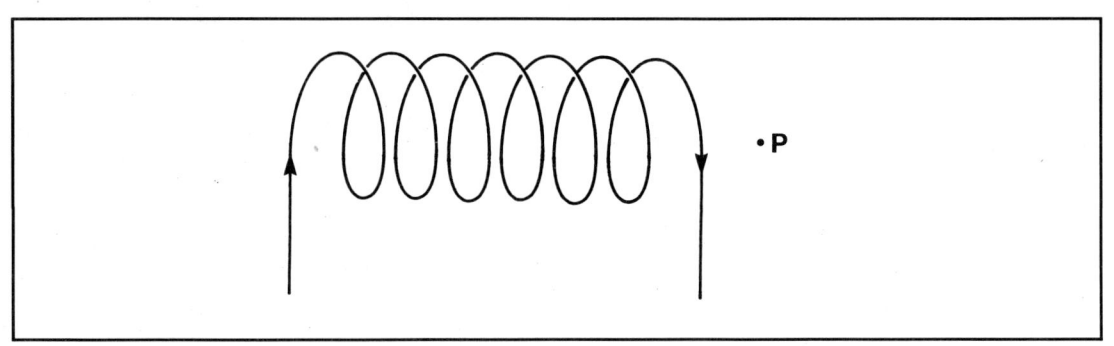

The direction of the needle of a compass placed at P when the current is flowing is:

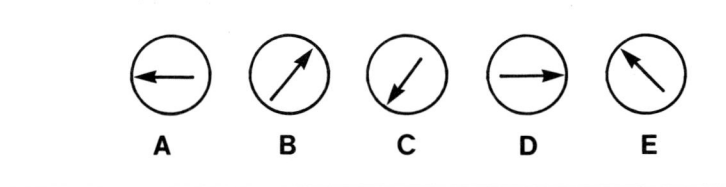

A ←   B ↗   C ↙   D →   E ↖

5. If you are connecting with a 3-core cable to a three pin plug, what colour must be connected to the live terminal?
   A — Black
   B — Brown
   C — Blue
   D — Green
   E — Green and Yellow

Questions 6-10

For each of the questions ONE or MORE of the answers are correct. Decide which of the responses is (are) correct. Then choose

A	B	C	D	E
1,2,3 correct	1,2 only	2,3 only	1 only	3 only

6. When a current flows through a wire, which of the following effects does it produce?
   1. Heating
   2. Magnetic
   3. Chemical

7. If an electric current flows through water with a little acid added, it produces
   1. Hydrogen
   2. Oxygen
   3. Nitrogen

8. The job of a fuse is to:
   1. Allow current to flow to earth
   2. Switch off the current in a circuit if it becomes too large
   3. Melt if the current is too large for the circuit   □

9. Which of the following fuse/fuses **could not** be used with a 2KW heater connected to a 240V supply?
   1. 3A
   2. 5A   □
   3. 13A

10. Electromagnets are used in:
    1. Electric motors
    2. Scrapyards   □
    3. Electric bells

Questions 11-15

In the following questions you must decide if they are TRUE or FALSE. If true put T, if false put F.

11. A wire carrying a current has an electrical field around it.   □

12. A solenoid is a long coil of wire.   □

13. 1 kilowatt-hour is the amount of electrical energy converted to
    other forms in one hour when the rate of working is 1KW.   □

14. Watts = Amps/Volts.   □

15. A 2KW fire working for 2 hours uses 1KW of electricty.   □

Structured Section

1. The three-pin plug shown is wired for use on a 240V, 1KW electric heater

a) Complete the table describing the wires X,Y,Z.

Wire	Name	Colour
X	earth	
Y		
Z		

b) The plug when used is placed in the wall socket shown below. Mark on the diagram the 'hole' which is connected to earth, using the letter E.

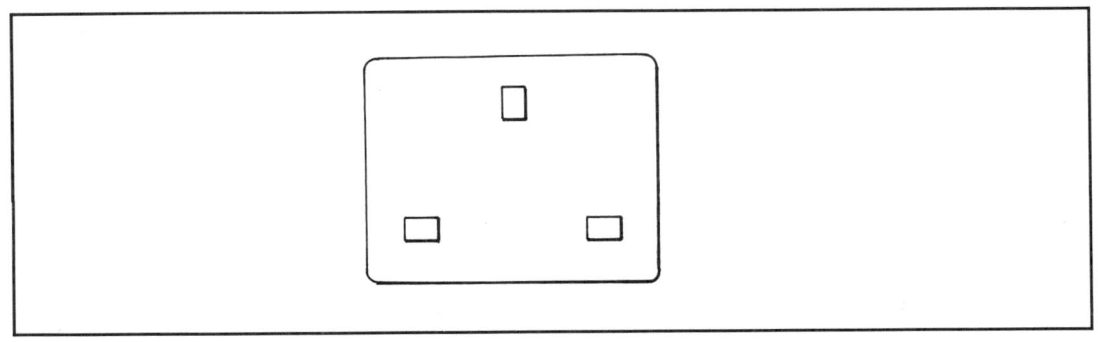

c) To which part of the electric heater will the earth wire be connected? ....
....

d) What is the reason for this connection? ....
....

2. The diagram shows two wires passing through a card which are carrying equal currents. The magnetic field around one of the wires has been drawn. Draw in the other.

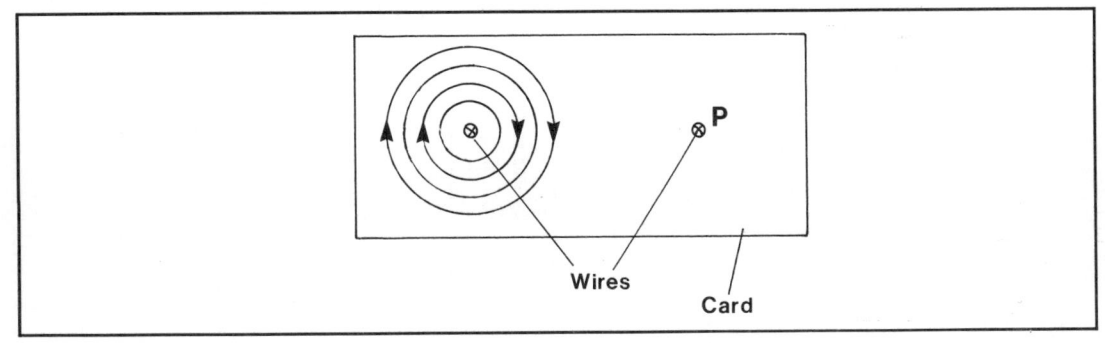

a) Where the fields overlap, does the resulting field become stronger or weaker? ..
....

b) Draw in the space below a diagram to show the effect of reversing the current in P.

Essay Section

1. What is meant by an electromagnet? Describe two situations where it is used.
2. Give a detailed description of an experiment to show the magnetic field around a straight wire.

Practical Assessment

Aim: To demonstrate the chemical effect of an electric current

Apparatus:

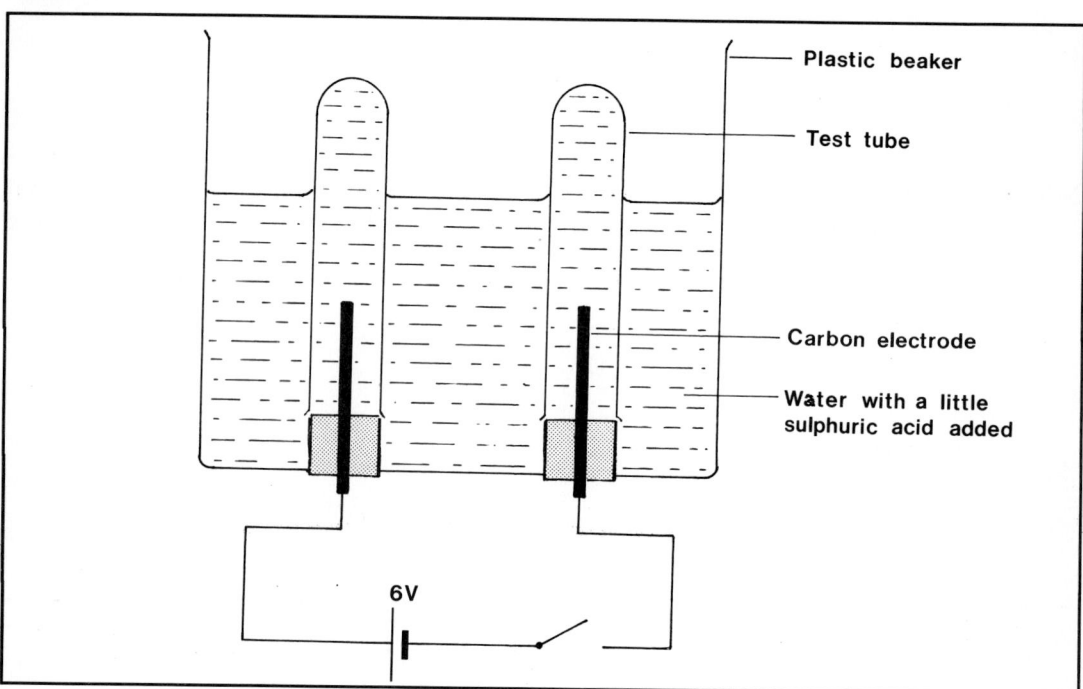

Method

1. Collect all the apparatus and then connect it together as shown, but leave the

battery disconnected.
2. Get your teacher to check your circuit and tell him/her how you will connect in the battery.
3. Fill the test-tubes with the acidulated water and then place them over the electrodes as shown.
4. Connect the battery into the circuit and switch on.
5. When one of the test-tubes is full of gas, open the switch. Look carefully at the connections to see which terminal of the battery is connected to this electrode.
6. Remove the test-tube, keeping it upside down and hold a lighted splint to the mouth. Note what happens.
7. Remove the other test-tube and place a glowing splint in it. Note the reaction.
8. Write up the experiment.

43 — Energy and Heating

Multiple Choice Section

Questions 1-5

Each question is followed by five possible answers A,B,C,D and E. Choose the BEST answer.

1. The unit of specific heat capacity is:—
   - A — J/Kg
   - B — J/°C
   - C — kg/°C
   - D — Jkg/°C
   - E — J/(kg°C)

2. If 4,200 J are required to raise the temperature of 1kg of water by 1°C. How much energy is needed to raise the temperature of 2kg of water by 3°C?
   - A — 8,400 J
   - B — 12,600 J
   - C — 25,200 J
   - D — 2100 J
   - E — 700J

3.

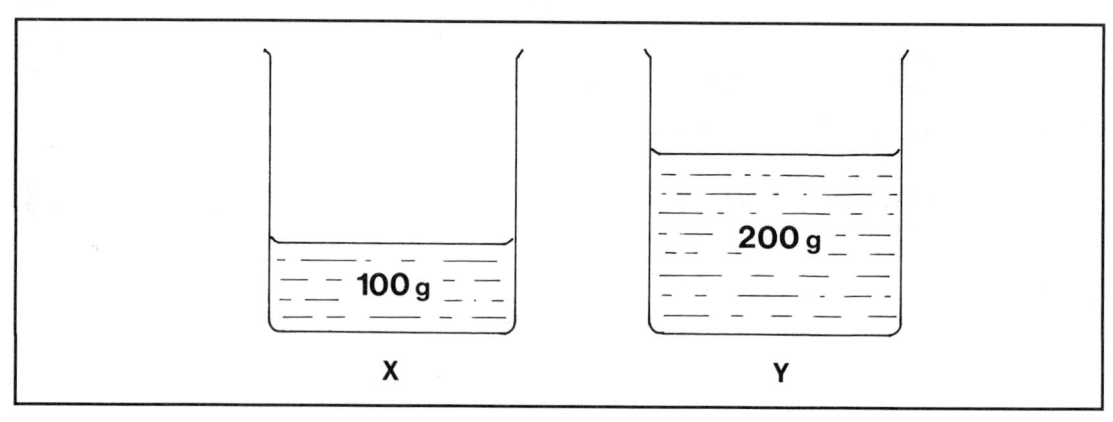

In the diagram above Y contains twice as much water as X. If equal amounts of energy are supplied to each beaker and contents, how would the temperature change?
   - A — X and Y change by the same amount
   - B — The change in X is double that of Y
   - C — The change in Y is double that of X
   - D — There is no change in temperature
   - E — The temperature of the beaker changes but that of the water doesn't

4. The unit of Specific Latent Heat is:—
   A — Jkg°C
   B — J/kg
   C — J/°C
   D — Jkg/°C
   E — Jkg □

5. Which of the following must be used to find the energy required to change a certain mass (in Kg) of water into steam?
   A — mass × sp. latent heat of fusion
   B — mass × sp. heat capacity
   C — (mass × sp. latent heat of vaporisation)/temperature change
   D — mass × sp. latent heat of vaporisation
   E — mass/sp. latent heat of fusion □

Questions 6-10

For each of the questions ONE or MORE of the answers are correct. Decide which of the responses is (are) correct. Then choose

A	B	C	D	E
1,2,3 correct	1,2 only	2,3 only	1 only	3 only

6. When a substance is heated its internal energy increases. Which of the following is increase?
   1. The kinetic energy of the molecules
   2. The potential energy of the molecules
   3. The number of molecules in the substance □

7. If energy is supplied to a liquid which of the following can occur?
   1. The temperature falls
   2. The temperature is constant
   3. The temperature rises □

8. The energy given out when a body cools can be calculated if which of the following are known?
   1. Mass of body
   2. Specific heat capacity
   3. Change in temperature □

9. Which of the following are important when finding the specific heat capacity using the method of mixtures?
   1. Heating the body in boiling water for 10 minutes
   2. Quickly transferring the body to the water
   3. Recording the highest temperature reached □

10. Which of the following graph/graphs show the temperature changes which could take place when a body is given energy?

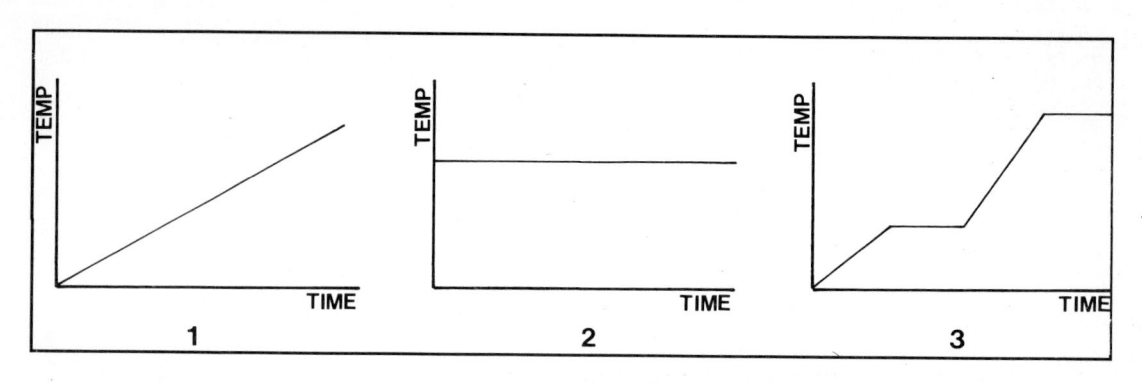

Questions 11-15

In the following questions you must decide if they are TRUE or FALSE. If true put T, if false put F.

11. A joule meter measures the energy supplied to a heater. ☐

12. Copper has a larger specific heat capacity than water. ☐

13. The specific latent heat of a substance is the amount of energy required to change the state of 1 kg of the substance. ☐

14. The equation for the energy required to raise the temperature of a mass m by $\emptyset°$ is $mc/\emptyset$.

15. When ice change to water, the temperature remains constant. ☐

Structured Section

1. In the method of mixtures 500g of Copper is heated to 100°C and then added to 200g of water. The final temperature of the mixture is 35.4°C.
   Specific heat capacity of copper 400J/kg C
   Specific heat capacity of water 4200J/kg C

   a) What is the mass of the copper in kg? ....

   b) How much energy does the copper lose when it cools from 100°C to 35.4°C? ..
   ....

   c) Give the mass of the water in kg. ....

   d) How much energy does the water gain? ....

   e) Given $E = M_w C_w (\emptyset_f - \emptyset_i)$. Find the initial temperature $\emptyset_i$, of the water ....
   ....

2. 3kg of ice at 0°C is changed into steam at 100°C.
   Specific latent heat of fusion = 334 KJ/kg
   Specific latent heat of vaporisation = 2240 KJ/kg
   Specific heat capacity of water = 4200 J/kg C

   Calculate the following

a) The energy required to change the ice to water at 0°C ....
....
b) The energy needed to change the temperature of the water from 0°C to 100°C .
c) The amount of energy required to change the water to steam. ....
....
d) What is the energy required to change the ice to steam? ....

Essay Section

1. How would you prove that the melting point of Napthalene is 80°C. Include a diagram of the apparatus in your explanation.
2. Describe a method for finding the specific heat capacity of a solid.

Practical Assessment

Aim: To find the specific heat capacity of water.

In order to find the specific heat capacity, 100g of water is heated. The temperature and the energy supplied are noted every 2 minutes. The results are given below.

Time/mins	Temperature of water/°C	Energy Supplied /KJ
0	20	0
2	55	15
4	79	30
6	92	45
8	100	60
10	100	75

Instructions

(a) Using the results above, plot a graph of temperature against time with temperature on the y-axis. Then answer the following questions.

(b) What is the temperature rise in 6 minutes? ....

(c) How much energy is supplied in this time? ....

(d) Using the information available to you calculate the specific heat capacity of the water ....
....

(e) Why is the graph a straight line after 8 minutes? ....
....

ISBN 1-85276-003-6